S0-BBI-538

ROSALIND FRANKLIN
and DNA

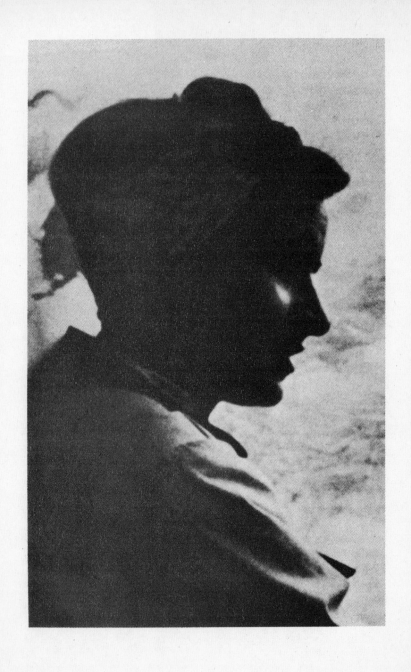

Anne Sayre

ROSALIND FRANKLIN and DNA

W · W · NORTON & COMPANY
New York · London

Cover photograph of Rosalind Franklin used with permission of
Contributions of 20th Century Women to Physics, copyright (CWP)
and The Regents of the University of California, 1995–1999.

Copyright © 1975 by Anne Sayre
First published as a Norton paperback 1987; reissued 2000
All rights reserved

Printed in the United States of America

Library of Congress Cataloging-in-Publication Data
Sayre, Anne.
        Rosalind Franklin and DNA.
        Includes bibliographical references.
        1. Franklin, Rosalind, 1920–1958. 2. Deoxyribonucleic acid. I. Title.
QP26.F68S29   1975   574.8'732'0924   [B]   75–11737

ISBN 0-393-32044-8

W. W. Norton & Company, Inc., 500 Fifth Avenue, New York, N.Y. 10110
www.wwnorton.com

W. W. Norton & Company Ltd., 10 Coptic Street, London WC1A 1PU

1 2 3 4 5 6 7 8 9 0

In memory of

A. LINDO PATTERSON
1902–1966

AND

ISIDOR FANKUCHEN
1904–1964

Contents

Acknowledgments

I suspect that few people before in the whole history of the printed book have owed so much to so many people as I do now. In no way at all could I have written what I have written without the cooperation and tireless assistance of all those who told me what I did not know, explained what I did not always easily understand, and readily provided me with encouragement at every point.

The list of those to whom I owe gratitude begins with my husband, David Sayre, who has been from start to finish inexpressibly helpful and understanding. He has patiently expounded the science which appears in this book; he has guided, taught, and corrected; he has lived uncomplainingly for nearly five years with my labors, doubts, distractions, and distresses; he has been at times my loving adversary as well as my constant strong support, and surely this is remarkable. This book would not have been possible without him.

From Muriel Franklin, Rosalind's mother, I have had every conceivable assistance and very great kindness and generosity. Dr. Aaron Klug has not only supplied a vast quantity of necessary material, but has given me much patient explanation and valuable criticism.

Dr. Delia Agar, Mrs. Simon Altmann, Dr. Geoffrey Brown, Dr. William Cochran, Mrs. Francis Crick, Dr. Jean Hanson, Dr. Philip Hemily and Marian Hemily, Dr. Peter Hirsch, Dr. Dorothy Hodgkin, Dr. Kenneth Holmes, Mrs. Gertrude Horton, Dr. André Lwoff, Mrs. Anthony North, Mrs. Anne Piper, Sir John Randall, Dr. Harry Carlisle, and a number of people who prefer not to be named have all contributed greatly to my information and understanding through their willingness to talk to me at length about Rosalind Franklin and her work. From Rosalind's good friends Vittorio Luzzati, the late Denise Luzzati, the late Jacques Méring, Rachel Glaeser, and Mme. Adrienne Weill I have received invaluable help. Dr. Raymond Gosling has given me considerable insight not only into the work upon which he collaborated with Rosalind at King's College, but into the complex relationships which prevailed there at the time. Dr. Robert Olby has generously shared some of the material from his forthcoming book on the history of the discovery of the structure of DNA. To the sensitive perception of Dr. Mair Livingstone I owe an understanding of Rosalind's last years which I could not have received from any other source.

Professor R. G. W. Norrish, Dr. Katarina Kranjc, Dr. Drago Grdenič, Miss Anita Rimel, and Mrs. Irene Neuner have been generous and informative correspondents. Especial gratitude is owed to Dr. Maurice H. F. Wilkins and Dr. James D. Watson, both of whom consented freely to lengthy and frank interviews. To Dr. Francis Crick my indebtedness is particularly great, and no thanks I can offer is adequate.

From Dr. Gabrielle Donnay, Dr. José Donnay, Dr. Carolyn Cohen, and in the highest degree Dr. D. L. D. Caspar, I have received encouragement and assistance which has made this book not only a reasonable undertaking, but a possible one.

This book was originally begun in collaboration with Dr. Einar Flint. When he was unable to continue in a joint effort, he provided not only lasting interest and encouragement, but all the material which he had independently amassed, for which I am deeply grateful.

Where I have erred, no one is to be blamed but myself. But where I have not, it is to these people that the credit is owing.

The President and Fellows of Lucy Cavendish College, Cambridge, honored me with an appointment as Visiting Scholar; the Warden and Fellows of All Souls College, Oxford, hospitably allowed me to write much of this book under their ancient roof. Both of these institutions have my lasting appreciation.

Where I have neglected to acknowledge the thanks I owe,

I apologize. Certainly I cannot omit from this list of bene-
factors my publishers, W. W. Norton and Company, from
whom I have received much sustaining encouragement during
this very long project. To Elizabeth Janeway my warmest
thanks are due for her most helpful criticisms of my text. And
not only for friendly support, but for indispensably useful analysis of my argument, I should in particular like to thank James
Buchet.

ROSALIND FRANKLIN
and DNA

ONE

An Introduction

In 1968, a book appeared which was read with great interest and much pleasure by a large number of people. This is not surprising, for it was a sprightly and colorful account of the way in which an immensely important discovery in science was made, written by a Nobel Prize winner who was himself one of the makers of that discovery. The book was *The Double Helix,* the author was James D. Watson,[1] and what he had to tell was the story behind the establishing of the molecular structure of deoxyribonucleic acid. That substance, better known as DNA, is the source of genetic inheritance for all living things.

This was a story worth telling, and worth listening to. If nothing else, it would be difficult to exaggerate the importance of the discovery which Watson was concerned with, for it is one that comes close to being incomparable. Upon it rest many of the most significant and potent advances in biological science made in the past twenty years; the possibilities for investigation and further discovery which it opened up have by no means been exhausted. As a consequence of this discovery, a new science, molecular genetics, has developed, which has proceeded vastly to expand our knowledge of what can be called the design of life itself. Out of this accumulation of new knowledge and new insights, still further, and stunning, poten-

tials may now be glimpsed which could, in time and for better
or for worse, enable us to make fundamental and deliberate
alterations in that design. Such prospects have never confronted previous generations, and if they are now at hand, it is
because of events which culminated in the publication in 1953
of the structure of DNA. Few moments in history can have
been as consequential as the one which opened the door upon
this present and upon such a future; few stories can have as
much real substance to them. But it also mattered that the
story Watson had to tell was that of a participant: the inside
story, in fact, of how important scientific work is done, told
not remotely or at second hand, but by one of the doers.

Surely this was in principle an admirable endeavor. It was
an unusual one as well. Very little has been written by
scientists, and even less by scientists of great distinction, with
the object of conveying to the laity either the nature of research or the attitudes toward the work of those who do it.
This is one of the great, and greatly serious, failures in communication of our time. It is a failure which may one day produce bitter results. Upon what science can accomplish, and is
accomplishing, rests much of the determination of the shape
of the future; because of science, we may either live longer or
die sooner; because of what science can do, our children or our
grandchildren may inhabit a world that is at present almost
unimaginable, living in circumstances which are barely describable, confronted by choices of which we can scarcely conceive.
A society which is ignorant of these determining developments
until the moment when new discoveries take it by surprise is
trusting more to sheer good luck than can ever be recommended as a policy. The public has the right to know, and the
duty to ask; scientists have the responsibility of telling; and
that clouds of mystery have been allowed to gather around
research and those who do it is deplorable. The laboratory is
not Merlin's tower, and scientists are neither magicians nor
seers, but people not unlike the rest of us who are occupied
with matters that are complicated, but not in the least arcane.

It is good and necessary to remember this, but it is a subject
of which we are not often reminded. For a man like Watson,

deeply and very successfully engaged in the work of science, to write about an important discovery with a cheerful lack of inhibitions was most unusual, but this was exactly what he did. And so it might be assumed that we had the truth at last, from the best possible source, and for once presented without either mystification or condescension.

The Double Helix was, indeed, a cheerfully uninhibited book, and entirely free from condescension. Its object was to deal not with the technical aspects of scientific research in particular or in general, but with the way in which science is done, using one example of the doing which the author was careful to say was not necessarily representative.[2] Written as a kind of memoir, frank and chatty and sometimes gossipy, it provided one man's view of the world of science and its inhabitants, and sometimes offered surprises. Science, it told us, is done by ambitious and competitive people, and there is no use in quarreling with this statement, for all that it often appears to be done with more joy, happy curiosity, and sheer pleasure in the doing than Watson allowed for, and with rather less awareness, too, of what the people next door are up to. Never mind. The author offered a qualified truth, and announced the qualification frankly, for his book was subtitled, *A Personal Account of the Discovery of the Structure of DNA.*

To write a personal account of anything is to claim a certain license; one is not insisting, "this is how things were," but only, "this is how things seemed to me"; and in his preface Watson claimed just such license, pointing out that other participants in the story might well choose to tell it differently.[3] The personal approach, then, removes some of the burden which normally rests upon the responsible historian, while disarming those who, with a different notion of the story in mind, might be inclined to argue or to criticize. But even so, surely some question remains concerning the extent to which the license conferred upon an author who chooses to write a purely personal account of real events can legitimately be extended. If in these circumstances the opinions offered by the "I" who tells the story must be taken at face value, for the "I" is the core of the tale, then what about the facts?

Facts have something of a life of their own. They are by no means entirely subject to viewpoint: What one likes, or does not like, does not affect what *is*. Facts may be annoying, they may hamper the flow of a good story, or even contradict it; but when they are swept under the rug in order to let the tale get on, they remain facts, locatable, discernible, stubborn, and there for the seeking. In one instance, and in my mind, a question arose concerning the accuracy of some of Watson's facts, simply because he presented in *The Double Helix* a character named "Rosy" who represented, but did not really coincide with, a woman named Rosalind Franklin whom I had known, admired, and liked very much. What was questionable was not a matter of opinion. No person exists concerning whom all opinions are unanimous and identical, and of all things which cannot profitably be disputed, taste in personalities leads the list. If Watson's opinions of "Rosy" made Rosalind Franklin's friends indignant or resentful, that was too bad, but it was neither remediable nor requiring of a remedy. But there seemed some reason to insist, at the very least, that the "Rosy" whom Watson disliked should have been recognizable as the real woman she stood for, particularly as that real woman died in 1958 and could not defend herself against anyone's misrepresentations.

"Rosy" was not recognizable as Rosalind Franklin. She was recognizable as something else not related to the facts.

"Rosy" was less an individual than a character, in the exact sense, that is, of a character in a work of fiction. That novelists create characters we all know; it is their stock in trade, their method; it is what we admire them for when they do it well. But novelists also write novels, and plainly label them as such. *The Double Helix* was not a work of fiction, but an account—albeit personal—of certain events in history in which a number of real people had been concerned. And of these people, one who was no longer living had been curiously transformed into what she had not been in life.

Both the method of the transformation and its results were clear enough. The technique used to change Rosalind Franklin into "Rosy" was subtle, but really not unfamiliar; part of it, at the simplest level, was the device of the nickname itself, one that was never used by any friend of Rosalind's, and certainly by no one to her face. But beyond this, a series of minor inaccuracies of fact, each aligned in a consistent direction, and accompanied by an unvarying emphasis in minor details, served to form a character out of a real individual by a means not materially different from the one a novelist uses in creating a character out of the whole cloth. More important than how it was done was the result achieved. If Rosalind was concealed, the figure which emerged was plain enough. She was one we have all met before, not often in the flesh, but constantly in a certain kind of social mythology. She was the perfect, unadulterated stereotype of the unattractive, dowdy, rigid, aggressive, overbearing, steely, "unfeminine" bluestocking, the female grotesque we have all been taught either to fear or to despise.

Whether this grotesque exists anywhere in real life, or is truly a figment of myth and imagination, is beside the point. Certainly she did not exist in Rosalind. By masterly technique, which deserves demonstrating, a magical change was made that transformed the real into the fictive. Watson introduced his "Rosy" early in his book, in a long descriptive passage remarkable for the number of small errors of fact which it embraced, and for the negativism of all its characterizing phrases. It is here that we are told, at the outset, a true fact, which is that Rosalind was working in 1951 in a laboratory at King's College (London), and then, at once, an untrue one, which is that this laboratory was under the direction of Maurice Wilkins. We are confronted, that is, with a trivial error that appears barely worth correcting.

But wait. The stage has been set, and with great economy, for we have been informed, in this very easy way, that Rosalind was working *for* Wilkins; we are even assured that her presence in that laboratory was intended only to speed up *Wilkins's* research. The relative positions of two people have been defined, and with what follows, the "Rosy" character instantly emerges. For Rosalind claimed, Watson says—though whether on good grounds or no grounds at all he does not mention—

that she had been given the DNA problem for her own, and so refused to "think of herself as Maurice's assistant." [4] In the circumstances he has sketched for us, such a refusal can only have been uppish, mutinous. Who on earth, we are led to ask, does that lab assistant think she is? And so we accept, barely noticing what it is that we have accepted, a perfectly clear picture of an aggressive, perhaps belligerent, female subordinate with no respect for her superiors, and an unlovable tendency to get above herself.

This is very fine art, to be so persuasive so briefly. But the picture offered rests upon air; the facts contradict it. The organization at King's College in which both Rosalind and Maurice Wilkins worked was under the direction of Professor (now Sir John) Randall, and in this laboratory they were on a level of equality, each supervising a group—in Rosalind's case, a very small one—engaged in entirely different research activities, though both were concerned with DNA. The DNA problem, if it belonged to anyone, belonged to Randall's laboratory. One aspect of the research being done upon it was certainly Rosalind's responsibility, for this approach made use of X-ray diffraction methods, and this was a technique which she commanded that Wilkins at the time did not. She had, in fact, been brought into Randall's organization in the first place in order to organize, supervise, and carry out that aspect of the DNA work, while Wilkins specialized in biochemical and biophysical studies. And suddenly it appears that the young woman who refused to think of herself as Maurice Wilkins's assistant was quite justified in her reluctance, for this she never was, nor was ever meant to be.

We have been misled. Facts dissolve our faith in "Rosy." It is possible that Watson was also misled, and assumed a situation at King's College very different from the one which prevailed, but at no time can it have been difficult to discover the facts, which were never concealed, or unclear, or stated otherwise by either Randall or Wilkins. What people assume provides as good a basis as any for forming judgments about their intentions, and Watson's assumptions may fairly be taken as indicative. For in the same long passage he gives a description of his "Rosy" that seems also purposeful. "Rosy" is something of a termagant; mere inspection by Watson indicates that she is willful and unbending. By choice, he tells us, she refuses to emphasize her feminine qualities—by which he means (and it is his idea of femininity) that she is badly dressed, wears no lipstick, does nothing interesting with her straight black hair. As "Rosy" is the model of the bluestocking, of course she peers out at the world from behind her spectacles, and we know she does because Watson muses upon how she might look if ever she removed them. (And those spectacles are, as it happens, rather crucial.) He is kind enough, however, to want to find excuses for her being as she is and, therefore, conjures up an unsatisfied mother hovering somewhere in the wings who has urged a professional career upon her daughter, hoping to save her from a dull marriage; he is surprised to find that this really cannot be the explanation for "Rosy's" defects, because no pushing mother can be located in a comfortable, prosperous background, and therefore regretfully concludes that "Rosy's" austere qualities are all of her own making.[5]

Now, all of this is thoroughly absurd. It is absurd in general. It can never be of much importance how any scientist, male or female, chooses to dress, and few of them of either sex are noted leaders of fashion; it can never matter how any scientist looks, and few of them are ever likely to be confused with film stars, nor is there any reason why they should be. But it is just as absurd in particular. To like or dislike anyone's style of dressing is a matter of taste. People with whom Rosalind Franklin worked in both England and in France thought her rather smart, always well-groomed, discernibly English in her style, but far from habitually dowdy. The unadorned lips that Watson found distressing simply indicate him to be unobservant—the lipstick was almost invariably there. And of all things, he cannot really have wondered often or long how Rosalind would look without spectacles, for he cannot often have seen her wearing any. Rosalind had the eyesight of an eagle, and resorted to magnifying lenses only for the closest of fine work.

None of this, of course, matters a particle in any very real sense. But what "Rosy" has been arranged to suggest matters quite a lot, indeed.

Elizabeth Janeway put her finger on it in commenting upon Watson's book in *Man's World, Woman's Place*. She wrote,

> We may, however, take advantage of his candor to note Watson's idea of where women belong in science: outside it. On the one hand we have Rosalind Franklin, a capable (if sometimes mistaken) research scientist in the King's College (London) team headed by Maurice Wilkins, which was working on the structure of the DNA molecule in competition with the Cambridge team of Watson and Crick. Watson's description of "Rosy" is personal and cruel. He is, of course, personal about everyone, and everyone is first-named, but no one in the book is so constant a target for aggressive attack as Rosy. She dressed badly, was stubborn in her views, harried her boss, wore her hair unbecomingly—in every way she was unsatisfactory, save as being the villainness of the piece. . . . Introducing her, Watson writes, "The real problem was Rosy. The thought could not be avoided that the best home for a feminist is in another person's lab." Clearly Rosy, a normally good scientist, is abnormal as a woman.[6]

There is a great temptation to believe that it was for the purpose of making this suggestion that "Rosy" was devised. It is not one that the personality of Rosalind Franklin would have supported.

Very well; let us assume that this is so. It is then surprising that Watson, who has a remarkable gift for candor, resorted to such obliquity. If he is among those who feel that the intelligent use of intelligent women's intellectual gifts and aptitudes is an affront to the natural order of things, he might have made the statement quite openly. It is not—unfortunately—so unusual, so original, or so generally unacceptable a view that it would not find as many to support it as to deny it. It is a view of woman's place so traditional that only recently, and with much effort, has it become available to argument, attack, and (possibly) defeat. But it is possible that "Rosy" came into being not just as the wicked fairy in his tale, to illustrate a notion of bright women in general, but for other reasons as well.

That is what this book is about. Here those illusory spectacles perched on Rosalind's nose become important. When *The Double Helix* appeared, I read it as innocently as any member of the lay public might, eager for enlightenment. That I was acquainted with a number of the people Watson mentioned provided no particular insights into his narrative—I am not a scientist, I am only married to one, and my relations with scientists never rest on a professional basis. We find other subjects to talk about, or we do not talk at all. I had no information with which to challenge Watson's narrative, which, for all I knew, might well have been accurate and objective to the smallest detail, and deeply imbued with that sense of fair play which he claims in his preface.[7] That I disagreed with his opinions concerning Rosalind Franklin was proof of nothing except that Watson and I liked different people. But those spectacles were not a matter of opinion. They had been provided for "Rosy," of my own knowledge I knew that they did not belong to Rosalind. There is an old maxim with which lawyers are familiar that warns one to be alert for any departure from known fact in a witness's testimony, for any such break opens for scrutiny all other facts to which that witness has testified. And so I have scrutinized, and this book is the result.

This is in no sense a biography of Rosalind Franklin. Biography is too cruel a word to use in connection with a life which was over long before it was finished. Rosalind died at the age of thirty-seven, and this can only be described as too short a life. Everything which ought to be present to fill the last chapters—the continuing successes, the mounting rewards—is missing. Her work went on almost to the day of her death, with unfailing brilliance and productivity; her rewards were few. That they were approaching when she died is beyond doubt.

The New York *Times* in reporting her death praised her as one of "a select band of pioneers"; [8] had she lived until 1962, the Nobel Prize Committee in considering its awards would have been required to confront the legitimacy of her claim to a share in the honors divided among Watson, Francis Crick, and Maurice Wilkins. For all her life she went on from strength to strength, and there is no telling where her own powers might have taken her. But none of this can be recorded, only supposed.

The value of her contribution to the discovery of the structure of DNA is not, I think, disputed by anyone. Owing to a curious set of circumstances, it was probably more significant than she herself realized, and it is partly to make clear how these circumstances arose, and what they resulted in, that this book has been written. The period during which Rosalind was working on DNA was a rather unhappy one in her life, and one during which she had cause to feel the disadvantages of being a woman working in a world where women were—and still are—somewhat unusual. This was not always the case with her; at other times and in other places she did little complaining, plainly because she found little to complain of. Her demands were reasonable and rational because she was a reasonable and rational person, and when she found that they were not satisfied by the situation she occupied at King's College (London), she moved elsewhere, was contented, and put past episodes behind her. There they remained while she lived; it was only the publication of *The Double Helix*, appearing ten years after her death, which resurrected them, and then in a fashion very little to her credit. It is doubtful if Watson's book could have been issued in the form in which it was while Rosalind was alive, for she was not at all a meek soul, and she would have made her protests heard.

The protests I am making here on her behalf are not necessarily the ones she would have chosen to make for herself. But there is, indeed, another side to the DNA story, and it is Rosalind's.

TWO

Rosalind

People who knew Rosalind Franklin well describe her differently, but not so differently that their reports bewilder. There is, if anything, a fundamental agreement in which the same words often recur, and are varied chiefly by a difference in emphasis. This is not to say that everyone who knew Rosalind interprets her character in the same way, for that would be too much to expect. Interpretation reflects the interpreter, who cannot help but supply, consciously or otherwise, the responses elicited. Rosalind's was a reserved nature, chary of self-comment. It was also a transparent and consistent one that reserve did not make in the least opaque.

The words people call upon to describe her are strong and vivid ones that do not convey, and are not meant to convey, a pallid or uncertain personality. She was not a mouse; few people who met her, even briefly, can have failed to notice her. Rosalind had "presence," of the kind that made people think her taller than she was, and this did not come entirely from a straight carriage or the possession of vibrant physical energy. It did not come entirely from her striking good looks either, or the rather elegant, neat swiftness with which she habitually moved. Her quality may have been to some extent inborn; evidently it manifested itself early, for her mother has written,

> Rosalind felt passionately about many things, and on occasion she could be tempestuous. Her affections both in childhood, and in later life, were deep and strong and lasting, but she could never be demonstrative, or readily express her feelings in words. This combination of strong feeling, sensibility, and emotional reserve, often complicated by intense concentration on the matter of the moment . . . could provoke either stony silence or a storm. But when she was a child, frustration tended to produce vehement protest, with sudden angry tears. . . . These storms were as a rule quickly over, even if sometimes they were too easily provoked. But the strong will, and a certain imperiousness and tempestuousness of temper, remained characteristic all her life.[1]

The intensity was what everyone who met her recognized.

The story of her life, brief though it was, is in fact a passionate one. That what she did with her capacity for passion was to devote it to science is no odder a choice than any other would have been. Rosalind had in a very high degree the talents, the mental capacities, the kind of intelligence, which lend themselves happily to science: they are the ones which embrace a gift for logic, an endless and very specific curiosity, and the ability to take enormous and positive pleasure in the workings of reason. If it is the passion with which she used these endowments in doing science that seems odd, that is because scientists in general are often mistakenly thought to be far cooler than in fact they are. Science, like art, is too demanding to be done perfunctorily; only technicians can get by on detachment, diligence, and patience alone. The difference between Rosalind and other good scientists was never that she possessed an unusual passion that the rest lacked. It was, at the outside, a matter of degree; possibly, but in no way is it provable, she was more intense than most.

The consequence of this was a quality in her that is uniformly described by everyone who knew her as dedication. She was, and indisputably so, a dedicated scientist. It could be left at that, if the word did not have a slightly different connotation when applied to women, which is to say that to speak of a dedicated man conjures up an image not quite the same as mention of a dedicated woman. Dedication in a man suggests a priestlike quality, a willingness selflessly to serve that which is greater than self, and if this service demands sacrifices, they can be taken without further examination not only as respectable, but as noble. It is very rare for a man who is a scientist to be described as dedicated, though he may be called brilliant, a genius, a marvel at his work; and this is evidently because, in a male scientist, dedication is not called for. His work is his pleasure, his means of expressing himself, his arena for commendable achievement, and the notion of sacrifice connected with dedication is entirely absent. Science also qualifies nowadays, in social and economic terms, as a good job.

Rosalind's work was also her pleasure, her self-expression, her area of challenge and achievement, but it does not end there, for she was dedicated. And, indeed, she made sacrifices, of an order usually thought of as respectable, even noble, when made by men. Here the shaded meaning comes in. A woman who, for example, chooses not to marry is rarely thought of as having chosen so much as of having failed. We have long been assured by persistent rumor that priests lead fulfilled lives, but that nuns are forever to be suspected less of possessing fulfilling vocations than of having retreated from life into the convent. Dedication in a woman may be necessary, but it has the look of being not quite natural.

Whatever Rosalind had chosen to do, she would have brought to it a single-minded and single-hearted devotion of great intensity. Her choice was science. Her commitment to this choice was total; it was also joyful. Few people ever can have taken more pure delight in what they did with their time and their talents and their energies than Rosalind took in her work. If this is included in the term *dedication*, certainly it was an entirely positive one, rewarding, triumphant, never dreary, and never a substitute for something else that was lacking.

But a better description, less confused and compromised, is

that Rosalind possessed a fierce but happy sense of vocation. It is true that this capacity is not often found in women, but one cannot easily argue from its usual absence that women are peculiarly disqualified from having it, if only because there have been just enough women, from Joan of Arc onward, who have exhibited it to provide evidence of the opposite. If it appears infrequently, that is because it is something women are talked out of, usually early in life, by persuasive and sometimes forceful arguments. Women's education has not generally been directed toward developing a high sense of vocation, nor have girls generally been encouraged first to seek out their talents, and then to cultivate them unwaveringly.

There is possibly an aspect of kindness in this discouragement, for a deep sense of vocation is no guarantee at all that the discovered and developed talent, if possessed by a woman, will find the opportunity to express itself; the chance of sheer frustration has always been considerable. It has been safer to encourage girls who are interested in medicine to take up nursing than to urge upon them hopes of someday becoming doctors, and there have always been good reasons for telling young women intent upon business careers to improve their typing. These cautions may well reflect nothing more than a practical recognition of realities, but a sense of vocation must be unusually powerful to survive them. If it is to flourish undiscouraged, a few favorable supporting circumstances are required. To this extent, Rosalind was born lucky, for she had better than average opportunities to identify and cultivate her talents, and what was on the whole a very favorable situation for the development of her capacity for commitment.

She might even be said to have inherited it, if not in the strictly genetic sense, then in the sense of a pervasive family tradition. She was the offspring of two remarkable strains that had produced, through a number of generations, individuals of marked ability and distinction—the Waleys on her mother's side, and on her father's, the Franklins—and this cannot be assumed to have happened entirely by accident. Concerning the Franklins, a genealogically inclined journalist remarked that "no student of Anglo-Jewish communal history will have diffi-

culty in assessing the benevolent potency of heredity as exemplified in successive generations." [2] He was not exaggerating
the evidence from which his conclusions were drawn. The significance of the family history lies in the persistence of character. From the first Franklin to settle in England down to
Rosalind, the character was distinct.

The first Franklin * to come to England was Abraham, son
of the Rabbi of Breslau, who settled in London in 1763, where
he taught for a time before establishing himself successfully in
business. He was a learned man, evidently energetic, who left
behind him a reputation for an unusual devotion to charitable
activities which his descendants proceeded faithfully to maintain and enlarge upon. The family business was merchant
banking, and it prospered; Rosalind's great-grandfather, son of
the first Abraham, was able to keep his carriage, and to occupy
a large house in Porchester Terrace, a handsome part of London, that was handed on to his son Arthur Franklin, and which
remained in the family's possession until it was bombed beyond repair in the Second World War. To this Arthur Franklin
added a country house in Buckinghamshire. So a picture of
considerable worldly success emerges, but this was never anything like the whole story.

The family character dominated. It included a great capacity for vigorous commitment, often exerted, and sometimes
very originally, in the direction of philanthropy. As orthodox
Jews, the Franklins accepted the obligation to be charitable,
but the way of their charity was energetic and far-ranging.
They were doers, in contemporary terms, natural activists.
They devised intelligent schemes and carried them out with
vigor as well as devotion; and these covered a list of social
projects so formidable and varied that there seems no end
either to their labors or to their devising. From working-class
housing schemes to the organization of schools of religious education for the children of poor Jewish parents, from baby-
care institutes in the slums to—eventually—the causes of wom-
en's suffrage and socialism, the interests represented by one

---

* The family name was originally Frankel, early anglicized to
Franklin.

Franklin or another ranged everywhere. The intensity of personal involvement was often very great, and it did not stop with the men of the family; the Franklin women worked beside the men, when they were not running ahead into projects and causes of their own.

They all possessed great determination. Rosalind's grandmother, Caroline (Jacob) Franklin, was a woman of remarkable accomplishments. She was an educationalist of some authority, who managed board schools in the East End slums under the London School Board; she sat for thirty-three years as one of the two women members of the Buckinghamshire Educational Committee, filling a public post; she founded and supervised a Jewish Lads' Club in East London, and carried on in the same depressed area a welfare center for mothers and children which had been organized by her own mother, Julia Jacob, while finding time to serve on a committee dedicated to the rehabilitation of those women tactfully known at the time as "unfortunate." * She was also the mother of six children, and one of them, her daughter Helen, has written about her: "She was a wonderful mother, never letting her public work stand in the way of the welfare of her children. We were never the kind of children who saw our mother for a few odd hours in the evening. We went everywhere with her."[3]

This was not a commonplace woman. Some of her work still endures, though much of it has since passed out of the hands of charitable individuals into the functions of the state. That it was done at all testifies to a capacity for intense commitment. The example was faithfully followed by her children in their various ways. Three of her children became active socialists, abandoning the family's Liberal political tradition, partly on

---

* A record remains of a speech made by Caroline Franklin in connection with this work which delights because, two generations later and in another context, Rosalind might have said the same words: "The only work that is degrading or derogatory is bad work, work that is done solely for what it will bring, and not for the sake of the work itself."

the grounds that socialism seemed a logical extension of concern for others, but also partly in the interests of women's rights. The oldest daughter, Alice, became a socialist at an early age, joining a group of young intellectuals headed by H. G. Wells, and subsequently worked for all her life on programs of social reform, especially women's causes. Arthur Franklin was proud of his wife's varied activities, but his views concerning his daughters were conservative; "Alice and I," Helen has written, "said that he recognized three sexes: men, women, and mother." [4] Alice did not go to college, though the ban was later lifted for Helen. Of Rosalind's aunts and uncles, Alice was the one to whom she was closest; they had a lasting friendship that existed quite outside the blood tie, and which exerted at times some influence upon Rosalind.

Helen also became a socialist, an active women's suffrage worker, a trade union organizer, and in her middle years had a political career as an elected member of the London County Council, of which she was for one term the chairman—a position barely describable in American terms, but in English ones, comparable to that of the Lord Mayor of London. These are unusual activities for a woman who could reflect, "I became ashamed of taking my friends home to what I considered . . . unnecessary opulence, although I must confess I always enjoyed the comfort in which I lived." [5] Quite evidently, opulence did not corrupt a committed conscience, nor can it have had any effect upon the third socialist in the family, Rosalind's uncle Hugh, who was converted to advocacy of votes for women while at Cambridge, and embraced the cause with such fervor that he went to prison for six weeks in 1910 for having made a violent attack with a dog whip upon the person of Winston Churchill, then a prominent antisuffragist. Hugh's dedication to his chosen cause was fiery; it was also far too well publicized for the comfort of his family, owing to the fact that he was the nephew of the Home Secretary of the time.*

---

* Arthur Franklin's sister Beatrice was married to Herbert Samuel, later High Commissioner for Palestine under the British Mandate, and later still, Viscount Samuel.

He married a fellow suffrage worker, Elsie Duval, who died not long afterward, and it was for her that Rosalind was given her second—and never used—name of Elsie.

The capacity for passionate dedication operated steadily; it was plain enough in Rosalind's father, Ellis Franklin, whose zeal in pursuing his commitments was different in direction but in no way less than that of his socialist sisters and brother. It is surprising that in a family where opinions differed so radically, and passions habitually ran high, divisiveness was absent. Disagreement led to argument, but not to ostracism; hot tempers were displayed, but not cold shoulders. Arguments might go on and on, inconclusively, for years and years, bursting out with fresh fire at any opportunity, but there is no sign that smoldering and unconfessed resentments interfered with what were generally comfortable and affectionate family relationships. What is not surprising is that Rosalind, with this tradition behind her, often behaved in accordance with it, sometimes to the bewilderment of those who did not understand how it worked, and assumed that the only outcome of a sharp, hotly argued disagreement was either hostility or capitulation.

This is one sort of family history; there was, of course, a more intimate one. When Rosalind was born on July 25, 1920, she was the second child and first daughter of a very happily married young couple prepared to welcome not only a new baby, and a girl, but the three additional children who came along in the course of time. Both parents were very young, under twenty-five at the time of Rosalind's birth, but they had already established the pattern of a contented and distinctive domestic life. Happy households communicate their quality and their standards much more strongly than unhappy ones, and certainly Rosalind came from a home that was, by all objective standards, close to ideal. Its character was rooted in an exceptionally successful marriage. That it was this we know because Ellis and Muriel Franklin described themselves as "supremely happy," [6] and proved the innocent assertion over a long number of years. Tolstoi was wrong when he began *Anna Karenina* with the curious statement that happy families are all alike, for experience proves the contrary. Human misery knows only a few repetitive forms, but what makes for human happiness is so remarkably diverse that any family which exhibits it demands precise description. The one that produced Rosalind was both a model of matrimony and an example of domestic idealism, and all the evidence indicates that on neither score could it be faulted.

It represented a fusion of the Franklin strain with another line of inheritance equally distinguished for its production of remarkable individuals, though the gifts they displayed were not identical ones. Muriel Franklin was born a Waley. Waleys and Franklins had been acquainted for more than a hundred and fifty years, since the time when Waley children had been sent to learn their Hebrew and their Law from Abraham Franklin. The families were allied by marriage as well, so that Muriel and Ellis had cousins in common, though no actual blood relationship. The Waley traits ran to the intellectual; they were cooler, on the whole, but no less determined, than the fiercer Franklins. Jacob Waley, Rosalind's great-grandfather, was the son of an East India merchant, and was one of three children, possessing a brother who was a brilliant composer and pianist, and a sister who had sufficient talent as a painter to be accepted as a pupil by David Cox and Copley Fielding. Jacob himself went to University College (London) at the age of thirteen, holding a scholarship in mathematics. When the University of London was empowered to grant degrees, he became its first graduate, taking first-class honors in classics and mathematics, and, indeed, he won a gold medal as a mathematics prize. He subsequently became a very successful and admired barrister, and while practicing actively at the bar, held a professorship in political economy at University College (London), thus becoming the first Jewish professor in an English university. In connection with his appointment as counsel to the Court of Chancery, he wrote a book on conveyancing which is still in use today. And while engaging in this busy dual career, he was one of the founders of the Jewish Board of Guardians (today the Jewish Welfare Board).

Jacob Waley was a man of great culture and genuine brilliance. He married Julia Salaman, a niece of Sir Moses Montefiore, and sister of Sir David Salaman, the first Jewish Lord Mayor of London, and the first Jew to take his seat in Parliament.* The marriage, which was a very happy one, produced six children; one of them, Rosalind's great aunt Cissie, was a social worker who left her mark as a founder of both the Union of Jewish Women and the Education Aid Office—the last an organization formed to assist worthy but poor young people to the financial support they needed for professional education. Rosalind was considered to resemble Cissie in personality, if not in looks. Another Waley daughter, Rachel, was the mother of the poet and Orientalist, Arthur Waley.† A persistent gift for the arts, sometimes for literary composition, sometimes for music, and sometimes for painting, ran through several Waley generations, and made an appearance in Rosalind, at least to the extent that she was capable of turning out, at an age when she still signed her name in uncertain block letters, a charming sketch in paint-box water colors of a landscape in which, astonishingly, the perspective was correct.

An alliance between Waley and Franklin was plainly that of equals as far as heritage went; the marriage between Muriel Waley and Ellis Franklin possessed another equality, which was that implicit between two people who had everything in common, and just enough difference in temperament to provide a comfortable domestic balance. On values, standards, objectives, and every matter of significance, they agreed profoundly, and if their approaches or methods sometimes differed, this was both appropriate and harmonious. One of the things they shared was a deep and clear notion of the central

---

* Sir David Salaman was elected to Parliament three times; on the first two occasions he was denied his seat because he refused to take his oath upon the New Testament; the third time his rights were recognized, and the test oath which had hitherto disqualified Jews was abolished.

† Rachel Waley married David Schloss; the family name was changed at the outbreak of the First World War, as happened in a number of families whose surnames had a German connotation. The Schloss connection was a double one, as Muriel Franklin's mother was born a Schloss.

importance of domestic life, and they had their own concept
of how it should be led. This was a simpler one than had usually prevailed among well-to-do people of a previous generation, and though it included comfort, it turned away firmly
from formality. The atmosphere cultivated was a kind, cheerful, busy, affectionate one that embraced the presence of the
children, not in nursery isolation at the top of the house, but
as active participants in family doings. The children were also
regarded, and from birth, as distinct and individual personalities who were to be encouraged to cultivate rather than to
repress their individuality in wholesome and productive ways,
an idea rather commoner today than it was fifty years ago,
and revolutionary compared with the rigid practices of child-
rearing which had preceded it.

This is, indeed, a sketch of a good place in which to grow
up, and the one possible defect of concentration upon the life
within the home, which might have been a certain narrowness
or isolation from the world, was avoided because social conscience forbade it. Ellis Franklin, even when a very young
husband and father not long back from the war, had his commitments. He devoted what came close to a lifetime of hard
voluntary work to the Working Men's College, an institution
founded in 1854 by Frederick Denison Maurice as an outcrop-
ping of the Christian Socialist movement. It aimed at bringing
together men of the working classes and men from the universities with the common aim of teaching and learning; Ellis began as a teacher, and later became vice-principal. The war, and
his early marriage, had prevented him from carrying out an intention to study science at Oxford or Cambridge, but he had
acquired sufficient command of some technical subjects to become a rather notably successful teacher—today the physics
laboratory at the College bears his name. His devotion to this
institution did not entirely win Rosalind's approval, owing to
the exclusion of working women; but in the 1930s, when he
gave nearly all his time to the Center at Woburn House organized to aid in the rescue of Jews from Nazi Germany, they
saw so much eye to eye on the matter that she worked there
for a time herself, as a schoolgirl volunteer.

It was one of her few excursions into the world of social action so dear to her kinfolk. What Rosalind derived from them was not an interest in reform, but a capacity for deep commitment that could operate in any chosen direction. She inherited intelligence and energy; determination and dedication are presumably communicated in other ways. In her case, they were chiefly communicated by reformers, idealists who were far from ineffective, because they were not inclined to let difficulties deter them from what they conceived to be right action. The list of these is long. Because of them, Jews sit in Parliament, poor children were for years aided to the education their talents deserved, mothers learned to rear healthier babies, learning was spread among those whose opportunities to receive it were limited, social outcasts were rehabilitated, the threatened were led out of danger into safety.

Granted that for many of the reforms in which they were so active the historical moment had arrived; nevertheless, each one came to fruition when it did because there were a few individuals who were determined not to let the moment pass unnoticed. A certain kind of vision was required, and a certain kind of persistent determination too; zeal was needed, but also a great deal of common sense. What communicated itself strongly and clearly might be called an idea of fixed values, inherently if not always explicitly moral, but worthy of a single-minded personal devotion, even to the point of dedication. In Rosalind this turned itself to science. There was no other way she could imagine of doing it.

Such sincerity can be as unnerving in the laboratory as elsewhere. In science, too, an affectation is sometimes practiced that is calculated to charm. It is that pretence of taking part in a complex but frivolous game in which work must always appear to have been modestly, and disarmingly, tossed off without effort, by sheer chance, but also presumably out of the natural ebullience of genius. What charms is the apparent lack of self-importance, of taking oneself too seriously; but what is implied is that the work itself is trivial, not important enough to require application or devotion. In any profession, there is often some friction between those who practice this artful ca-

sualness and those who do not. Rosalind did not practice it.
Her attitude toward her work was deeply serious, and toward
its demands she was uncompromising. Considering what she
had early learned at home from numerous examples, it could
not be otherwise. If she had immense, and sometimes formidable, determination, that was because there was something
important to be determined about.

The difficulty with determination is, of course, that to those
who do not share its purpose, it sometimes seems quite indistinguishable from intractable stubbornness. It is admirable
when we agree with its object, and merely pig-headed when
we do not. It is a characteristic which also is likely to lead to
argument, and Rosalind, certainly, rarely flinched from a good
argument. This suggested to milder temperaments more tem-
pestuousness than was necessarily there. What was there was
a commitment, absolute and profound.

A total, undistracted, and undistractible commitment can
serve to focus talent, or so we are obliquely told, and very
often, by those who reproach working women for allowing
their multiple commitments to interfere with their concentration upon the job. In Rosalind's case, the question never arose.
She was single-minded. She had a capacity for tact, but she
was also extremely honest, and if tact and honesty conflicted
on any important matter, the honesty won. Because her own
commitment was total, she was scornful of triflers in science,
an attitude which might have been a harsh one if she had not
been entirely free of the peculiar snobbery of the talented,
who sometimes exhibit a tendency to disdain those whose gifts
are less than their own. Science is rather prone to its own form
of snobbery because it has something of a crude caste system
built into it. The aristocrats who are theoreticians look down a
little upon the solid middle class of experimentalists and re-
searchers, who in turn snub the proletariat of simple scientific
workers. Neither this nor any other hierarchy held the least appeal for Rosalind. She was unusually capable of respect for those
unlucky enough to find themselves doing uninteresting or unsatisfying work, and she could be gentle toward those whose
capacities were limited, provided that their hearts were pure,

and their efforts equal to their abilities. But what struck her as slipshod, lazy, careless, or unthorough got little sympathy, and often less tact. She was to that extent demanding.

There is no mystery about the origin of these qualities. Rosalind exacted a great deal from herself, not because she was exalted by notions of self-sacrifice, but because the standards to which she had been exposed, and with which she never quarreled, required this as a necessity. She could stick to her guns with immovable stubbornness when she thought she was right because no matter of importance could be lightly abandoned; that would have been betrayal. In commenting upon Ellis Franklin's long service to the Working Men's College, James Laver wrote, "Some people thought he had too much influence and that he always wanted his own way. Perhaps he did: he usually got it, and, in the end, he was nearly always right." Laver, who worked with Ellis Franklin for a number of years, listed his qualities: "strength of purpose, complete unselfishness, utter devotion to what he conceived to be his duty." [7] It is not a bad description of Rosalind.

It is clear enough where Rosalind's character came from. If she was not born with it, then she developed it very early. She had her father's will and fierce determination, but she had another side as well, for in her happiest moments she was plainly her mother's child—gay, literally sparkling, brimming with a slightly teasing, mischievous wit. These two aspects even had a physical expression, for in the first of them, Rosalind somewhat resembled the Franklins, and in the second she looked rather like the photographs of Muriel as a young woman. These were not conflicting elements of personality, they merely alternated according to circumstance. Both evidenced themselves in childhood, and never disappeared.

Certainly Rosalind had a happy childhood, but its particular implications are worth observing. As one of five children, she was neither smothered nor ignored, but until her sister was born eight years later than Rosalind, she lived entirely surrounded by brothers. This meant that in the normal competitiveness of close family life, her challengers were boys, and no more than any other girl well-endowed with brothers did she develop any exaggerated awe for the superior capacities of males. Equally, her brothers were companions and allies with whom she shared a good many activities, games, and common experiences. Because the Franklin children grew up in a cohesive family, and in a sustaining domestic atmosphere, they acquired the peculiar and identifiable confidence possessed by those who are part of a proud and compact group that cherishes its own standards, its own viewpoint, its own strong loyalties, and this confidence communicated itself to Rosalind as much as it did to her brothers. The strength that she drew from this background she took for granted, but it provided her, nonetheless, with certain advantages. If she was sometimes shy, she was never distressed or weakened by fundamental self-doubts.

This, too, had its consequences. From her earliest life, Rosalind was sensitive to opposition or frustration, and responded to both with fierce and stubborn indignation. She was not often opposed or thwarted, and when she was, there were usually reasonable grounds for it; but on no occasion did she take it really meekly. When she was very young, Rosalind suffered a severe infection which was not, in those preantibiotic days, easily curable; it recurred once; and for some time afterward it left her rather frail and susceptible to fatigue. No more than any young child was she capable of understanding a regimen designed to prevent overexertion, and she very much resented rests and naps and mild restrictions on her activity, the more acutely because she believed that her brothers, who were strong and vigorous, had no such impositions placed upon them. For a child not by nature languid, no doubt, even necessary caution was a grief; to Rosalind, it became something of a grievance as well. Despite the general egalitarianism prevailing among the Franklin children, or possibly because of it, Rosalind retained from her childhood a distinct impression of having been placed, with respect to her brothers, at an artificial disadvantage based upon nothing but sex, and much of this feeling seems to have originated during this fairly brief time when the real difference, if any, was no more than a matter of health.

Of course, it was unreasonable, but so children are. Her

sensitivity to her "special" situation might have been less if a degree of companionship with her brothers unusual for its time and place had not generally prevailed. To this day it is possible, and not extraordinary, for a well-brought-up English girl to remain cloistered in an almost exclusively feminine atmosphere until she is close to marriageable age; forty years ago, this was not only the possible, but the very probable, pattern. It did not occur in Rosalind's upbringing. When Ellis Franklin bought a carpenter's workbench in order that the children might learn some useful skills, the children included Rosalind, who probably profited most of all of them from learning to dovetail and miter. She was, anyway, unusually adept in later life at the machine-shop work that plays a certain part in scientific research. Being, in general, very little excluded from what the boys were up to, she took any exclusion hard, no matter what occasioned it—to her, the principles of preventive medicine were doubtless far less clear than the presence of restrictions which applied to her alone.

This early experience of illness had other effects. It left Rosalind with a lasting impatience toward illness of any kind, which she preferred to disregard insofar as possible, accepted only with intense resentment, and—rather puritanically— tended to consider a sign of personal weakness. She developed a physical stoicism which sometimes proved extreme. On one occasion during the war, she knelt on a sewing needle, drove it deep into the knee joint, and then walked, alone, and for a fair distance, to a hospital where she could have it removed. The doctor who attended her was flabbergasted, pointing out that she had done the impossible, that no one could endure to walk with a needle angled across the joint. Rosalind laughed at him. But possibly the strongest consequence of brief and early ill-health was that it set firmly her pattern of reacting to frustration with passionate indignation, of fighting every inch of the way rather than submitting for a moment.

It never left her. Nor did the suspicion then acquired that it was disadvantageous to be a girl.

How much, if any, immediate reality gave birth to this suspicion? This is extremely difficult, probably impossible, to

determine; it could never be determined absolutely, because
what is dealt with here are such delicate shadings of psychological outlook that even those who expressed them may well
have been unaware of them. Certainly Rosalind was a loved
child, and by no means an unnoticed one. She was carefully
reared, not only in the protective sense, but with much attention paid to her particular qualities.

Her education was excellent. She was sent to St. Paul's Girls'
School, the younger sister of that ancient foundation, St. Paul's
School, and the choice was a suitable one for her. It is true
that Franklin girls had gone there before, her aunt Helen and
a number of her cousins among them, and other cousins were
her contemporaries there; but whether tradition played a part
or not, it is hard to think of a school more suited to Rosalind
herself. Its standards were very high. Much was expected of
its pupils, which may account for the fact that a considerable
number of them have gone on to notable careers. Even forty
years ago it gave courses in physics and chemistry which were
very nearly as good as any offered in any school in England,
and rather better than those available at the time in, say, a
small American college. Its attitude toward its students was
essentially serious, and no other would have satisfied Rosalind.
It cannot be said that because she was a girl, she was less dis-
cerningly educated than her brothers, and indeed, she never
claimed this.

What can have been so troubling that as a grown woman she
would sometimes refer to her youth as a period made tense,
at the very least, by her need to struggle for minimal recognition?

Not, surely, the irksomeness of prescriptions for conserving
the strength of a small, frail child. There must have come a
time before Rosalind was very old when she began to view the
contents of that tiresome experience in a reasonable light, and
if she went on referring to it, this must have been for what it
symbolized to her rather than what it was. It may have had
exaggerated importance to her because she was sent away from
home to seek a cure at a boarding school on the coast which
specialized in the care of frail children, and was very homesick

and forlorn until she was brought back at the end of the year: cured, incidentally, and never to be frail again. But very likely it functioned actively as a symbol because it was something concrete, whereas most of what disturbed her otherwise was far less definable, no more than a matter of attitudes.

Attitudes may be implicit as well as explicit, but only the latter can effectively be argued with. In any practical or realistic sense, Rosalind confronted no opposition to her plans or hopes or ambitions sufficiently strong to require serious consideration. She made up her mind when she was fifteen concerning what she intended to do with her life, and she carried out her intention without any deflections at all, which is proof enough that no one opposed her except on a rather theoretical level. This is not to say that there was no argument.

Ellis Franklin was frankly doubtful of the general utility, or wisdom, of professional education for girls; if he admitted exceptions, it is not plain that he considered his daughter to be one of them. On his side of the debate, there were some objectively valid points. Rosalind was under no necessity to earn a living, and could not be regarded as one of those girls for whom a profession provided the best assurance of a secure future, for her family could, and did, provide for her quite adequately. Professions have not always been rewarding for women either—if there were in England, in the 1930s, some few women working seriously and professionally at science, they had not made much of a visible mark. None of these women scientists had yet been elected a Fellow of the Royal Society, none as yet held a major university appointment.* To

---

* In March 1944, Kathleen Lonsdale and Marjory Stephenson were elected Fellows of the Royal Society, the first women to be so honored. Concerning this, Dame Kathleen wrote, "Since then . . . there has usually been no more than one woman elected each year as compared with some twenty-five to thirty-five men. . . . Not even the most ardent feminist, however, could claim that this is due to sex discrimination. The fact is that there are not many women in the top ranks of research and so qualified for nomination to Fellowship" ("Women in Science: Reminiscences and Reflections," *Impact of Science on Society 205*, no. 1 [1970]).

embark on a career with such limited prospects and such possibilities of ultimate frustration was not lightly to be considered. Was there, indeed, any point in it at all? Ellis Franklin recognized Rosalind's intellectual powers, and was proud of them, but that they should lead her inevitably into a working career in science was by no means as clear to him as to his daughter. He had an alternative suggestion: that Rosalind should take up—earnestly, with sincere application—some form of the voluntary social work which in the 1930s still formed an important, even essential, contribution to the general welfare, which needed devoted and intelligent workers, which had in the past provided and was still providing deep satisfaction to women of Rosalind's background, which was a central part of his own life.

The suggestion was well-intended. It reflected Ellis Franklin's own commitments and his own experience; it was not trifling. With respect to Rosalind, it was at best wide of the mark. Apart from the fact that one cannot always easily implant one's own enthusiasms out of enthusiasm alone, it did not recognize Rosalind's unusual quality. Her intellectual gifts were highly specific, and of an order badly fitted to social work at any level, simply because she had by nature the scientist's mind, which is uncomfortable in dealing with any problems to which there are only approximate answers. Not only would she have been uneasy in dealing with the uncertain human equations that are the basis of all social work, but she would also have had no talent for it whatever. Her heart might have been in the right place, but her mind was not. Rosalind's imagination was of another order entirely, incapable of being fired by general questions, but highly suggestible to subjects demanding precise analysis and tight chains of logical reasoning. Any form of social work, even at the most remote administrative levels, is fundamentally intuitive in its judgments, and Rosalind's intuitions were all governed by more abstract conditions. If social problems could arouse her sympathy, that was the limit of what they could arouse; the available methods of attacking them, in terms of voluntary work, were too vague to her, too personal, too logically unsatisfying. Interest often follows confidence, and certainly she lacked confidence; probably she was right in guessing that in this sphere she had nothing to contribute but her time, that her abilities would be not only wasted, but even something of a handicap. But it is also true that on this point she was more unpersuadable than reasonable doubt accounts for. She resisted.

She resisted strongly. She never entirely forgot that she had been confronted with an alternative that, from her point of view, required resistance. She was very young at the time, and it is quite possible that she put more passion into resisting than was objectively necessary, and fought harder in order to do what she wanted than any active or concrete opposition warranted. But it is easier to see her objections as excessively strong if one has never been young, female, gifted, ambitious, and wholly uncertain of the limits to which female gifts can successfully be carried. And probably Rosalind was correct in believing that a career of voluntary work would not have been urged upon her brothers, for example. What she did not see so clearly was that the implicit attitude probably had less to do with a personal situation than with society in general.

It can be argued on her behalf that she was very young and inexperienced, and also that the tendencies of society with respect to female talent were, in the 1930s, little analyzed. Few families brought up their girls exactly as they did their boys, for the simple reason that the difference in female and male destinies was so great, and so accepted, as to make the experiment a dangerous one. Boys grew up to take jobs, to make careers, to support families; their ambitions and aspirations required some consideration and encouragement because they were consequential; and into this pattern the ambitions and aspirations of girls fitted only very narrowly, never really in the direction of happy independence. In the circumstances, kindly and protective parents found it difficult to justify encouraging their daughters in what was likely to work out badly, and few of them did it. Because Rosalind grew up in a fairly outspoken family, was not snubbed when she spoke out, and rarely suffered the worst snub of all, which is the lofty adult refusal to argue with the child, she was that much the less prepared to have her most serious intentions taken less than seriously.

Of course no sensible adult takes the statements of the young concerning their own futures as necessarily prophetic. It is only by hindsight that the occasional wonder can be noted, the child who knew all the while what it was talking about. More often than not there is no wonder to observe. Enthusiasms prove temporary, and what the young and untested assume about their own talents, their own pertinacity, their own stability, may well be illusory; a concerned adult may well feel obliged to argue the case for doubt, prudence, and caution. And so it may have been. But Rosalind did not see it so. She reacted as bright and determined children do toward caution and prudence—with defiance.

What must be noted on her side of the case was that she had discovered young, by the time she was fifteen, the true framework of her future. She had no doubts about her intention to do science, and modern science is done in professional terms. The gentlemanly amateur of the eighteenth and early nineteenth centuries, with his telescope and his microscope and his charmingly detailed field observations, has virtually no contemporary counterpart, if only because private fortunes do not run to radio telescopes, electron microscopes, or computer installations sufficient to analyze the observations. No one educated in science in the twentieth century can conceive of it as anything but a full-time, and complex, professional undertaking. It was exactly this which Rosalind visualized. It was exactly this concerning which Ellis Franklin had doubts.

The gap between them was conceptual; certainly it had more than a little to do with the fact that she was a girl. Ellis Franklin, after all, had once conceived of a career in science for himself, and it is unlikely that he would have felt anything but pleasure at the prospect of a son picking up where he had left off. Because Rosalind was a girl, there was a reasonable side to his objections; because she was a girl, she saw them as unreasonable. Years later, long after she had established herself as a professional scientist of considerable reputation, she still referred at times to her early struggles, always stating these in terms of the special disadvantages which applied to daughters. Clearly the struggle was a real one to her, and clearly the sense of having struggled never left her.

But what the observer from a distance is conscious of is not the presence of active opposition so much as the lack of positive encouragement. Compared with other young women who have been determined upon a career, and who have contended not only with genuine opposition but deprivations, Rosalind had little to complain of. There was never any doubt of her receiving an education commensurate with her abilities; there was never any objection to her aspirations so powerful that she had either to abandon them, or break out of the situation which forbade her to follow them. Her struggle was largely a subjective one, less for opportunity than for positive approval. But there is no reason to despise its reality in her mind simply because it is not at all the same thing as the struggle which has often confronted the talented daughters of poorer families.

Talented girls born into disadvantageous circumstances often benefit from encouragement, and when they are applying their talents as a means of pulling themselves up out of their disadvantages, they rarely suffer from doubts. Nor are there many people who would have the temerity to tell a young woman with the ability to improve her status through professional education that she was better off refraining. Whether one is male or female, skilled labor brings higher wages than unskilled labor, and the working conditions in offices and laboratories are generally better than those found in mines or factories.

The need to earn a living is recognized as a legitimate spur to young ambitions. For women, it has been nearly the only legitimate spur, which is not quite the case with young men. A young man who can afford to do nothing may choose to be idle; but no one is surprised, or dismayed, if one of these fortunates chooses to go to the office daily, or to stand for Parliament, or confess a desire to be governor of New York or president of the United States. That a similar need for self-definition, self-expression, self-development, or self-fulfillment may exist in women is not often observed. In Rosalind's generation, it was nearly always not observed. In women, a great sense of vocation has been reserved, it appears, to those who needed one. Rosalind was never obliged to earn a living, and at every point in her life attractive, comfortable, and secure alternatives to an exacting career existed. If she ignored them, it was because she had a deep sense of vocation; but their very existence demanded of her an unusual seriousness, an unusual determination, an unusual passion.

The suspicion that she was unencouraged may well have served to strengthen her commitment—in a sense, she was volunteering to be a scientist in much the same way as so many of her kinfolk had volunteered themselves to demanding obligations that had nothing whatever to do with self-interest. But to remain a volunteer in difficult circumstances calls for considerable will or considerable stubbornness. Having determined to do what she was not encouraged to do, Rosalind set out to prove herself. When this intention is allied to great sincerity, great ability, and inflexible will, the combination can appear formidable.

Indeed, there was a formidable side to Rosalind. It was linked to a certain innocence. That both existed side by side is entirely reasonable. The innocence was partly that of inexperience, of a girl who had had a sheltered rearing in a close-knit family, who had been insulated from exposure to much worldly knowledge, who was, indeed, past twenty-one, and a university graduate, before she claimed an independence that went beyond independence of the mind. There was nothing abnormal, or even unusual, in the protectiveness which had surrounded her; very few English girls of her time and general background possessed, or dreamed of asserting, anything like the free-ranging freedom of their American counterparts; and, by comparison with many of her contemporaries, Rosalind enjoyed considerable liberty. But, nevertheless, it was strictly limited by conventions which she did not challenge. When she entered Cambridge in 1938, she was no more unworldly or naïve than most of the other women students, but she had, all the same, a deep and peculiar innocence that went beyond inexperience, and which to some degree she never wholly lost.

This other innocence was rooted in a kind of determined rationality, an unconquerable conviction that reason dominated, that sane and sensible people, anyway, preferred to act in accordance with logic and were, therefore, accessible to sound argument. If this is not an exact representation of how humans behave, it was a fairly exact representation of Rosalind herself; in the conviction of its general truth all her faith rested. Science is, of course, a rational business in which sound arguments tend to prevail, and the reasonableness of it suited Rosalind perfectly. Like many other scientists, she had a natural affinity for objective evidence and objective proof, and no natural ease with wholly subjective reasoning or purely approximate thinking. This is not to suggest that the scientific mind is rigid or narrow, for Rosalind, like other good scientists, possessed a knack for the less-advertised side of science in which hunches and guesses and an instinctive feel for the problem play more of a role than is generally acknowledged. But in matters where her instinct was less developed and her experience less, she held fast to a total and innocent faith in reason. Absurdities exasperated her. At the same time, the slow process of patiently leading recalcitrant thinkers to better thoughts by artful persuasions seemed to her a waste of time—a logical argument, cogently expressed, was surely sufficient to convince, and if it did not convince, then the case might well be hopeless and not worth pursuing. That intelligent people, plainly not hopeless cases, were sometimes proof against the strongest arguments of reason she could never quite believe, and any demonstration that this was so made her not only furiously impatient, but left her unreconciled to the reality.

This is a youthful belief, and an innocent one. Most people who are affected by it do not cherish it long; even a little experience invites cynicism. Rosalind never lost it. It survived a Cambridge education, though in the 1930s and 1940s there still remained evidence on every hand in Cambridge that very intelligent people were not in all things subject to the rule of reason. The university itself embraced a streak of irrationality which cannot have failed to affect, in some degree at least, nearly everyone who was exposed to it. This irrationality had to do with the higher education of women, and that this remained a nervous and unsettled question as late as 1938 is proof enough that sheer reason does not always convince, and certainly that it does not always convince quickly. The line of thought traceable behind policies concerning women's education as they slowly evolved is proof enough that history defies logical analysis, and that human behavior, even as exemplified in great intellectual institutions, does not always have much to do with what is reasonable.

The history of women's higher education in England often appears to be a saga of the conflict between idea and emotion, each winning alternate rounds in a battle that went on being indecisive for much longer than seems possible. There is no parallel to it in American educational history, though parallels might be found in other areas of American life; but then, there is no American parallel to either Oxford or Cambridge. In contrast to the multiplicity of institutions which have characterized American higher education, Oxford and Cambridge stood alone in England—Scotland and Ireland had their own universities— until the nineteenth century was well along. Their dominance was not only total, it was pervasive, for both had developed through six centuries or so certain functions not closely related to the education of the young. Both were centers of power, connected with the established church, with the civil power, and often, if often indirectly, with politics. They overshadowed rival institutions so thoroughly that no rival institutions flourished until the needs of the nineteenth century brought them into being; even then, the new universities grew up very much under the influence, and to a degree under the supervision, of the senior foundations. Higher education in general was inextricably entangled, in England, with the senior universities, and women's education was no less so.

To the idea of women's education, Cambridge responded— in slightly curious terms, it is true—by permitting the foundation of Girton College in 1869; almost simultaneously, Oxford accepted women on much the same terms. In both places women students quickly established beyond argument their receptiveness to education, acquitting themselves satisfactorily, and sometimes better than satisfactorily, in the standard examinations. But in neither place did the examination records, however pretty and irrefutable, especially convince those who wielded the university power that women really belonged within their ancient gates, and the educational privilege was grudgingly granted.

How grudging the welcome was can be judged by Virginia Woolf's remarks in *Three Guineas:* "At Cambridge," she wrote, "in the year 1939, the women's colleges—you will scarcely believe it, Sir, but once more it is the voice of fact that is speaking, not of fiction—the women's colleges are not allowed to be members of the university; and the number of educated men's daughters who are allowed to receive a university education is still strictly limited, though both sexes contribute to the university funds." [8] More than this, she went on, "the total number of students at recognized institutions for the higher education of women who are receiving instruction at the University or working in the University laboratories or museums" might not "at any time exceed five hundred, [though] the number of male students who were in residence in Cambridge in October 1935 was 5,328. Nor would there appear to be any limitation." [9] Virginia Woolf was saying that on this subject the university dithered, and that behind the dithering there was hardly a shred of discernible rational argument.

So it was. Part of the absurdity was that Cambridge allowed women only "titular" degrees, awarding these from 1922 onward, but withholding real ones—ones that were identical, that is, to men's degrees—until as late as 1947. The difference was a practical one. Bachelor's degrees awarded by Oxford and Cambridge have a significance beyond that of an ordinary diploma, for they carry with them the promise of a Master of Arts degree, which is not earned, and which confers the right to sit in the legislative body which ultimately controls university policy. In a rather astonishingly democratic way, both Oxford and Cambridge are fundamentally governed by their own graduates of mature years, who vote in their legislatures to elect new chancellors, to ratify the conferring of degrees, to deal with some financial matters, and formerly each of these bodies also sent to Parliament one member apiece, to represent each of the senior universities as separate political constit-

uencies. A "titular" degree did not signify the right to a vote or
a voice in the university government; even the women's colleges themselves, as functioning parts of the university, went
unrepresented in the body that effectively governed them.

The point of this is hard to see.[10] The effect of an inconsis-
tent policy, however, was quite visible. Upon the university
men, it was a poor one. In 1922, for instance, while the debate
over women's degrees was taking place in the University
Senate at Cambridge, men undergraduates made a violent attack upon Girton College, doing considerable damage before
the mob was controlled by the police. Not in such a manner are
the merits of ideas usually, or commendably, thrashed out in
intellectual circles. But the effect that it had upon university
women was possibly worse. Those who are subjected to consistently illogical treatment respond to it as best they can, and
this is usually with confusion. No matter how clearly the
women at Oxford and Cambridge could think upon abstract
subjects, they had very great difficulty in knowing what to
think of themselves.

There is evidence for this. In 1935, Dorothy Sayers published
a novel set in a women's college in Oxford—a place the author
knew well—which contains some interesting conversations that
illuminate this confusion. "The fact is," says a fictional Fellow
of the fictional college, an aggressively intellectual female, to
some of her more wavering colleagues, "the fact is . . . every-
body in this place has an inferiority complex about married
women and children. For all your talk about careers and independence, you all believe in your hearts that we ought to abase
ourselves before any woman who has fulfilled her animal
functions. . . . Look how delighted you all are when old
students get married! As if you were saying, 'Aha! education
doesn't unfit us for real life after all!' And when a brilliant
scholar throws away all her prospects to marry a curate, you
say perfunctorily, 'What a pity! But of course her own life
must come first.'" The reply to this, from the Dean of the
college, is, "I've never said such a thing . . . I always say
they're perfect *fools* to marry." [11] The theme is repeated, and
often; later the Dean complains that educated women are in a

cleft stick, for if they show ordinary human feelings, it can be pointed out that womanliness unfits them for learning, and if they don't, then it can be said that learning makes them unwomanly.[12]

The reasoning is nonsensical, of course; few people suggest that learning Greek or astrophysics necessarily makes a man heartless, and it is hard to see why the effect upon women would be so chilling. But much more than superficial confusion of mind is evident here. If this picture of intellectual women at a university written by one of their number can be taken for anything like accurate, and there is no reason to think that it is other than a fair representation of what went on in the 1930s, then the confusion was profound enough to have come close to splitting these poor women into conflicting halves: "woman" quarreling with "scholar" inside each of them, each half rather resenting and disliking the other, and each somewhat distrusting itself. No one's peace of mind can ever have been significantly advanced by such an internal situation, and yet, quite plainly, exactly this disturbing war was forced by circumstances upon all women who crossed into the university's precincts. They cannot really have enjoyed it.

Women very much like these were educating Rosalind when she was at Cambridge. They were endlessly, and helplessly, conducting the "either-or" debate—to marry or not to marry, to have a career or give it up in favor of matrimony—because this was a serious debate, an unresolved question, a set of values in dispute, a pair of alternatives that did not imply a spectrum of choice. If an occasional daring woman leaped over the debate, married, and announced her intention of having a career anyway, she could not always rely upon moral support for her brave resolution. If she held a fellowship in one of the women's colleges, she was expected to resign it upon marrying, though the men's colleges had stopped exacting this penalty for matrimony in the 1870s. It is true that she had the possibility of being re-elected in her new condition, but this was a possibility, not a certainty, for there were always those who believed that married women ought not to take the bread out of the mouths of unmarried ones, and others who simply disapproved of so cheerful an attempt to have one's cake and eat it too. The debate was, in short, a sincere one, seriously argued. That did not make it the less confused.

To this confusion, Rosalind responded logically. If the conflict implied was "either-or," and everything in her background as well as what existed in the university atmosphere kept assuring her that it was, then the logical answer was to make, once and for all, a clear choice. Irresolute arguments had no charm for her, nor did the notion of half-commitments. Quite clearly and consciously she chose the career, not the marriage.

It was, for her, a perfectly logical choice. She made it with determination, but contentedly. And with it behind her, it was clear to her that the question would now vanish, being resolved. All the world could easily observe that a serious choice had been made, a single-minded and single-hearted commitment undertaken, and all the world would be bound, quite simply, to respect it for what it was. Sex as a factor in one's working life would disappear, and nothing but sheer ability would count.

This was logical. Rosalind believed it. No more innocent conviction can exist. She believed it the more because her choice implied some sacrifice. Rosalind's view of marriage remained until she was in her thirties completely based upon what she had seen of her parents' marriage, which she took literally for a model and an invariable pattern: strong and dominant male, supportive female, and a home in which the wife, especially in her role as mother, was firmly, permanently, and wholly centered. This was so rooted in her that, indeed, no man who was not very strong in the way in which her father was strong ever attracted her. More than that, for men who were weak, submissive, pliable, unserious, unforthright, she had a lack of respect which at times came close to contempt. Marriage, contemplated in the only terms in which Rosalind could see it, was so opposed to the commitment she felt toward her work that to dismiss it as a possibility may have been something of a relief; but her love for children could be put aside only as a sacrifice.

Marriage and children were inextricably connected in her mind, and if marriage might, by some remote and doubtful chance, be reconcilable with her work, most certainly this did not apply to motherhood. A woman's place might not necessarily be in the home, but a mother's was, and to this Rosalind admitted no argument.* She loved children quite sincerely and simply. She enjoyed and appreciated them, and she had a natural understanding of them that seemed entirely spontaneous and effortless. The nieces and nephews and the children of friends to whom she was much attached—and who universally loved her in return—were second-best, the substitutes for the children she would have liked to have had. This was what she gave up as the token and sign of her sincerity and her commitment, and it is possibly because she felt the sacrifice deeply that she made the innocent assumption that rational people would easily understand without further demonstration that she deserved to be judged not as a woman scientist, but as a scientist pure and simple.

---

* There is evidence for these attitudes. One ambitious but not very gifted woman scientist was told by Rosalind, coolly but not unkindly, that she might be better advised to stay at home and look after her husband properly. Rosalind's private comment was, "There's no use in doing *two* things badly." When a very close friend of Rosalind's went back to doing medical research after the birth of her first child, Rosalind was distressed, and argued that she was "not being fair" to the baby, and that the presence of a baby necessarily pre-empted the time of even a very gifted woman. There is no reason to think that Rosalind would not have applied these standards to herself.

Rosalind and I toured nursery schools in Stockholm together in 1951. The beautifully appointed and modern premises filled with healthy and well-scrubbed children appalled her far more than they pleased her. She found the surroundings grievously impersonal, and insisted that no amount of professional care could substitute for "a mother being there." She was distressed that the staff who dealt with the children were all plainly professionals who did not even "pretend to act like mothers." That the children were wonderfully well-behaved did not impress Rosalind either; she shocked the director of one of the schools by remarking that the children were "like trained animals."

But there is not that much rationality in the world, and only an invincible innocence could imagine that there was.

It was by no means immediately recognized that Rosalind's passion for science amounted to dedication, or that this dedication removed her from the endless, brooding argument concerning what women were, or were not, meant to be. No doubt Cambridge was the wrong place to look for such compliance; but at least it can be said that Rosalind wasted no time in seeking it or demanding it or arguing for it. Her years in Cambridge were spent almost entirely in hard work, for she was also not going to waste the opportunities at hand. Cambridge was then, as now, a very great center of science; nowhere else in England was a better scientific education to be found than in the place in which the Cavendish Laboratory, under a succession of remarkable directors, had established an almost unparalleled tradition of research. Because the war broke out at the beginning of Rosalind's second year, neither the college life available to her nor her intellectual training was exactly what it might have been in quieter times. College life was, on the whole, dull. Most of the mild frivolities which enlivened it in normal times—dances, feasts, rags, picnics on the river— were pale imitations of the real thing, or were missing entirely. Rosalind wrote home that Newnham seemed to her "rather like a boarding school," [13] and in the circumstances this was probably just. The war, on the other hand, provided a certain element of disorganization both to the usual course of teaching and to daily existence.

The disorganization in the teaching of science was particularly noticeable, owing to the fact that a good many scientists had vanished into war research—"Practically the whole of the Cavendish have disappeared," Rosalind wrote [14]—and either were not present in Cambridge at all, or were there on a part-time basis. In some areas, the disorganization was severe: "Biochemistry," Rosalind reported, "was almost entirely run by Germans and may not survive." [15] This did not mean that undergraduates were necessarily the less well taught, for as a rule undergraduates have not much contact with scientists of the order who were fast disappearing, but it did mean that they were required to be rather more independent than usual because those who remained to teach were heavily burdened. This suited Rosalind very well. She was the next to youngest student in her year at Newnham, and her experience of personal independence was fairly slight, but her capacity to manage her own intellectual affairs was already well-established. She worked passionately—"8¼ hours a day in the lab" [16]—and thrived on it.

On the other hand, wartime conditions were distracting and sometimes fatiguing. "We do not so far have to carry gas masks," she wrote home, "but apparently we *do* have to spend hours in trenches every time there is a warning." [17] Warnings were fairly frequent, as Cambridge was encircled with military camps, especially R.A.F. installations, but it was never seriously attacked, although the German radio said otherwise. Concerning these reports, Rosalind wrote, "Do you ever listen to the news from Hamburg? If so you will realize that I was right to come to Newnham. . . . Girton was in flames last week. They repeatedly remind listeners that the R.A.F. makes Cambridge a military target, and many colleges have been in flames on various occasions. What do you think can be the object of putting such things in the news for *English* listeners who can see for themselves that they are false?" [18] But the interruptions irked her, and the more so because Rosalind, as a mild claustrophobe, loathed hiding in air raid shelters.[19]

The independence allowed her in her work she cherished. She was too independent, and too stubborn for one of her tutors, who predicted that she would come to a bad end; from Rosalind's point of view, he was merely among those men who did not like to be argued with by female students. The habit of working on her own initiative was established during these undergraduate years at Cambridge, and she never lost it. If she suffered in any way from a less-than-normal amount of supervision and guidance, it is hard to see how. The chief error that she made was one that is common to enthusiastic students, and she learned from it, though not until she had made it twice. At the end of her schooling at St. Paul's Girls' School, Rosalind had decided, for no very pressing reason, to push herself hard so that she could leave school a term earlier than was normal. Not much was gained by this, as in any case she could not go to Cambridge until the following autumn, but perhaps the sheer challenge attracted her. Her reward was six enjoyable weeks *en pension* in Paris, learning with other English schoolgirls to improve her accent in French, and her punishment was that she did somewhat less well in her examinations than she had hoped, or expected. In Cambridge she varied the error only to the extent of overworking so ardently and so exhaustingly in her last term that she took her final examinations in such a state of weariness and nervous tension that she failed of a first-class degree, and ended with a high second. Her disappointment was bitter. But the lesson was driven home, for though Rosalind remained an enthusiast for her work, and more than diligent, she never again succumbed to the temptation to wear herself out. Nor would she, when she had other people working under her supervision, allow them the luxury of overenthusiasm.

Rosalind took her relative failure hard. She had counted on a first-class degree, and not to have taken one was the first check she felt either to her ambitions or to her confidence. In fact, it was not in a practical sense important, for when she finished her undergraduate work, she was presented by Newnham with a research scholarship, and surely that would not have been forthcoming if the examination results had been literally interpreted. The research scholarship she very much wanted. It was the status that counted. To do graduate work she needed a college behind her; and in wartime, when the manpower and womanpower resources of Britain were closely directed, with the holders of science degrees particularly subject to direction, such sponsorship was more than ever necessary. But for all that Rosalind was set upon earning a graduate degree, she was pulled in two directions. Her supervisor, Professor R. G. W. Norrish,* has said that she was "obviously keen to take her part in the war effort," and this may well have been true. But that she was restless during that graduate year had another cause as well.

She did not get on well with Norrish. It is difficult to see

* R. G. W. Norrish (1897–   ) F.R.S., Sc.D., Ph.D. (Cantab.) Joint Nobel Laureate for Chemistry, 1967.

exactly where the friction lay. Rosalind was known to refer to Norrish later—unkindly, and not at all accurately—as an "ogre," but even if she saw him in this light, ogreishness in itself was never enough to cow her. For his part, Norrish saw her as highly intelligent, intellectually bright, and eager to make her way in scientific research, but he saw her, too, as "stubborn and difficult to supervise," and "as not easy to collaborate with." There is no reason to argue with him; Rosalind could be all these things. Norrish was also preoccupied with wartime research, which may well have limited their contact. One may guess at the root of the trouble between them—though it is only a guess—from a remark of Norrish's to the effect that Rosalind was not only dedicated to her work in science but passionately devoted to raising the status of her sex to "equality" with men. If this is how he saw it, then Rosalind's intransigent response may be taken for granted.[20]

The general notion of raising the status of women was never more than peripheral to Rosalind, and on the whole it irritated her for its imprecision. She was "feminist" only in the widest philosophical sense, not in an activist one. It is not always easy for men to see the difference, but to Rosalind it was a clear one. She was not engaging in any broad or sweeping challenge when she insisted that her own status be acknowledged as not only "equal" to that of any comparable male scientist, but to be quite indistinguishable as well, because to her the emphasis was solely upon *scientist,* not upon the adjective. She was not declaring war on behalf of women's rights, but demanding on behalf of science that those who served it be judged solely and wholly upon their abilities. For herself she asked no favors, privileges, or special, softer standards of judgment because she was a woman, which meant that, to her, equality was taken for granted. To have it raised, then, as a separate point was maddeningly illogical. All this was plain as day to her, and she never saw why it was, in so many other minds, so cloudy. If the point was ever raised between her and Norrish, or if she so much as sensed it in any attitude of his, her uncooperativeness was assured.

But he cannot have been expected to know this. Rosalind's

particular and firm attitude was unusual then; it is not much
more usual now. It was one that had little in common with
doctrinaire, or political, feminism, and for the simple reason
that it was not fundamentally feminist. Her attitude was one
of exacting professionalism, and what is most surprising about
it in Rosalind's case was that it existed in one so young, who
was barely at the beginning of a professional life. She worked
it out for herself, but she may have been reinforced in it at
this time by her friendship with Adrienne Weill, whose outlook was not very different. Certainly this friendship was important to her, and sustaining during the period of her difficulties with her professor. Two women rather than one may
have been confronting Norrish with their firm professional demands.

This is possible because Adrienne Weill was older than
Rosalind, and far more familiar with the professional woman's
outlook. Mme. Weill was a refugee from France who held a
fellowship at Newnham during the war years; she was in her
thirties at the time, and the mother of a young daughter; she
was also a scientist who later attained considerable distinction
for work in metallurgy. She came from a family of intellectuals, and was herself the daughter of a brilliant woman, an
articulate and active feminist well-known in France for publications that were well in advance of their times. Adrienne had
come to England at the last possible moment, having gotten
herself and her daughter to Bordeaux after the fall of France
to the Germans just in time to be included in the last flight of
refugees. When in England, she had benefited from the sponsorship of the agency in which Ellis Franklin was a leading
spirit. But insofar as her acquaintance with Rosalind was concerned, this was purely coincidental. She and Rosalind met at
Newnham, in a way which might have been awkward except
that Adrienne was both a sensible and an insightful woman
capable of discerning human quality when she saw it. From
an odd beginning, she and Rosalind became friends.

The odd beginning was related to Adrienne's fellowship at
Newnham, which was—curiously—partly supported by sub-
scriptions of a shilling a week volunteered by students. In re-

turn for this, the students had been promised the opportunity to have lessons in French conversation from this cultivated Frenchwoman, a pleasant enough *quid pro quo,* except that no one had informed Adrienne of the promise she was supposed to fulfill. The first she learned of it was when Rosalind knocked at her door to inquire politely when the lessons were to begin. Rosalind was interested in improving her French, and she took the promise quite seriously; for all that it came as a surprise to the teacher, Adrienne did not object to Rosalind's calm insistence that a bargain was a bargain. She gave the lessons to a solitary pupil, and discovered while she was doing so that Rosalind was worth the effort.

From Rosalind's point of view, Adrienne was something of a revelation. To begin with, she was entirely unaffected by British prejudices. If Adrienne was grateful to Newnham for its hospitality, she was not uncritical of what she found in a Cambridge women's college. The ambivalence of the university, and the women themselves, toward the matter of women's education puzzled her, and the notion of educating women only in order to waste their talents appalled her. Her own outlook on these matters was clear, calm, rational, sophisticated, confident, and purely French, and she could not help but communicate it to Rosalind, who thought much the same herself, but who had not before met a somewhat older woman, and a scientist, with such sympathetic views.

When Adrienne rented a house in Cambridge in order to take in student lodgers, Rosalind helped her to get it ready; in 1941, she moved there herself. This was Rosalind's first experience of living outside the family circle in a noninstitutional atmosphere, her first taste of unsupervised daily living, and she chose an agreeable place to try her wings. In Adrienne's house, the prevailing climate was lively, good-humored, a trifle chaotic, incessantly and bilingually conversational, politically left-wing, and distinctly intellectual—in the French sense, which is not at all like the Anglo-Saxon one. It is, for one thing, more combative; for another, it is more analytical and introspective, much less confined than British intellectualism to what is factual and objective. Rosalind's French ac-

cent improved under Adrienne's tutelage, but so did her powers of argument, and her ability to extend her reasoning into
areas which she had not previously explored.

Inevitably, Adrienne influenced her, but the influence was
that of support for what was already there rather than that of
introducing new elements into Rosalind's character. The two
had a good deal in common beyond the fact that they were
both scientists; Adrienne, like Rosalind, was a strong, vivid,
outspoken personality with a critical edge to her mind. She
was also warm and generous, and quite willing to give up the
advantage that superior age and experience provided her with
in favor of a friendship between equals. Rosalind responded,
but not to the point of surrendering independence of mind.
On politics, for example, Adrienne no doubt broadened Rosalind's tolerance toward the arguments of the left, but not to
the point where Rosalind stopped arguing in reply. Her fundamental impatience with political thinking was already established, and if she had moved away from her father's rather
traditional Liberalism to the ground occupied by the socialist
Franklins, that was as far as Rosalind was prepared to move.[21]

But certainly when Rosalind was restless and unhappy in
her situation as a graduate student, Adrienne's sympathy made
it easier for her to confront her exasperations, and to resolve
them by choosing to leave. A desire to do something more
closely connected with the war effort may have played some
part in her decision—it was 1942, after all; the war was distant from no one's mind—but in later years she never claimed
this as her reason. What she suggested was that she regarded
her year with Norrish as something of a personal defeat, and
one that she did not take lightly.

She cannot in any event have had many illusions about what
she might have done in a practical sense to aid the war effort, for
it was well known in university circles, at least, that women scientists in England (and for that matter, in the United States)
were not at all warmly encouraged to lend their skills in any
very active or direct way. Why this was so has never been satisfactorily explained, but the policy was clear, and the proof
that it existed rests in the striking absence of women's names

from the lists of the scientists who were engaged in war research, though this absence cannot be accounted for by lack of volunteers. In England, the likeliest war contribution available to a young female science graduate was a teaching job as a replacement for a man who had been directed into livelier things. A prospect of this order cannot have attracted Rosalind in the least. Her passion was for research, her interest in teaching was never more than slight, she was ardent to make her way, and the imparting of the elements of physics and chemistry to school children was not a promising path for an ambitious young physical chemist to follow. To give up her scholarship was, therefore, something of a risk for Rosalind, because once she had left Cambridge she would be subject to "direction," and concerning the place to which she was directed, or the work that she was given, her own preferences might not be consulted.

But here she was very fortunate. What she was directed into turned out to be her first career.

This is an approximate way of putting it, of course. Rosalind's career in science had, over all, considerable coherence. It can be seen as a logical and reasonable progression, interesting because it demonstrates so plainly that at the frontiers of research the neat subdivisions of science vanish; from this point of view it is not surprising that Rosalind began by working with coal, and ended by working with viruses. Nevertheless, her first reputation was founded upon her work with coal, and that it did not end there is testimony to the range of Rosalind's mind.

She was, however, very lucky at the outset. When she was just one month past twenty-two, she was appointed assistant research officer of the British Coal Utilization Research Association—BCURA, more pronounceably called CURA. As the name suggests, CURA was an industrial organization, but it was rather superior to the ordinary run of its kind. If many research associations sponsored by industry exist primarily for the purpose of promoting whatever products provide their support, CURA had higher intentions, and the war intensified them. Its first director, Dr. D. H. Bangham—with whom Rosalind was for a time to collaborate—organized a staff composed of "young physicists taken directly from the Universities," who were capable of "bringing with them a knowledge of the latest techniques developed in such leading schools as the Cavendish Laboratory," and though "scarcely any . . . had any previous knowledge of coal or coal combustion . . . such knowledge was held to be less important than the latest methods of scientific investigation." [22] This is CURA's own view of its history, but it is not mere boasting. The staff at CURA was very good, and Bangham saw to it that the study of coal was fundamental and far-reaching. It was far from a bad place for a young physical chemist to begin.

CURA was a very young organization, founded in 1938, and not yet in full operation at the outbreak of the war. The war had several effects upon it. One was that it expanded rapidly, both in terms of staff and of the variety of investigations it took on, for fuel was a war-related subject of considerable importance. Another was that its original premises were bombed, and that thereafter it led a scattered life in several locations in London and the surrounding suburbs. The youthfulness of the organization could also be stated in terms of the age of its staff, many of whom were not much older than Rosalind. Because it was new, and full of young people, and carrying on its work in small and shifing units, CURA fostered informality. It also allowed, by necessity, for a good deal of independence in research. No doubt there are disadvantages connected with carrying on research in wartime in a number of places not designed for the purpose and in an organizational framework which cannot be other than loose, but they must have been offset by some advantages, for CURA was from the start very productive.

If Rosalind can be taken as a measure of productiveness, the output at CURA must have been staggering. Independence in research suited her perfectly; she worked happily, and hard; between 1942 and 1946 had her name on five papers; she was the sole author of three of these. Each article represented a considerable, and sometimes a vast, amount of experimentation painstakingly carried out, and in Rosalind's case much of it was inventive.[23] She also wrote during this period the thesis which she submitted to Cambridge for her doctorate—awarded in 1945—but the name she was making for herself at CURA had far more significance than any academic title.

What she did was lasting. Most of Rosalind's work at CURA concerned the microstructures of coals, hitherto neither determined nor measured. The scale of the measurements involved is perhaps indicated by an experiment Rosalind devised in which the unit of measurement was a single molecule of helium. The importance of what she accomplished is unarguable, for her early papers are still fundamental references on coal structures, and are still referred to today. As for the nature of her work, in 1970 Professor Peter Hirsch at Oxford called it "remarkable. She brought order into a field which had previously been in chaos." [24] This is no small accomplishment for someone who does it all between the ages of twenty-two and twenty-six. The details of her work, which are immensely technical, do not matter for present purposes. What should be noted is that Rosalind, working as a physical chemist, for the first time began to explore the world in which, essentially, she remained for the rest of her life.

This is a world defined not by substance, but by dimension, the tiny world of objects of no more than a 50 Å size.* Rosalind discovered this subuniverse at CURA, and in its environs her attention remained centered. Here is the connecting link which makes all of her work into a logical progression, much more cohesive than would appear from a description of her as a scientist who worked first in physical chemistry, then in crystallography, and finally in molecular biology. But that this progression was possible, and natural, is something of a reflection of the development of science itself, pressing its explorations further and further into this tiny world which had only begun to be mapped in the present century. When Rosalind first began to survey this territory, its horizons were still limited.

---

* Å is the abbreviation for ångstrom unit, named for A. J. Ångstrom, a nineteenth-century Swedish physicist. Å = one ten-millionth of a millimeter.

Though studies of structures finer than that of the micro-structure of coal had been accomplished, the possibility of one day determining the infinitesimally small and highly complex inner structures of biological substances seemed remote. Neither she nor anyone else can have had this in mind except as a fairly distant objective. Within ten years the prospects were radically to alter, and Rosalind was among those who altered them.

At CURA, she found herself a woman launched upon a career that no longer had to be debated. Wartime in itself had put an end to the question of whether she should or should not work, and when the war was over, she was a young scientist with a reputation, confidently following prospects too good to be abandoned. What Rosalind considered the defeat of her graduate year at Cambridge was erased. She had her doctorate in hand, a list of important publications to her credit, and the comfortable assurance in her mind that she could work successfully inside an organization when sufficient independence was permitted her. All this had been provided by her association with CURA. Given the opportunity to prove herself, she had succeeded, and in the meantime, she had come to understand, though it came to her more as a confirmation than as a discovery, that what she liked to do she also loved to do.

Rosalind had also emerged, in these years, as a personality. Though she was still in many ways young for her age, some earlier tentativeness had vanished, along with the defensive-ness with which the young guard and disguise their uncer-tainties. She could still be defiant, but there was a light-heartedness to it that had previously been lacking. When, for example, the authorities at CURA forbade the use of the machine shop to any but specially certified personnel, and hung signs to this effect on the machines, Rosalind went on doing what machine shop work she chose—and she was very good at it—while cheerfully turning the signs blank side out-ward. An earlier Rosalind might have put up an argumenta-tive resistance. Nor would an earlier Rosalind have sewed brightly colored patch pockets on her lab coats because she was tired of their drabness. A new confidence had leavened her essential seriousness. It was supported by the fact that she and her father had come to the end of their debate over her future. She was a working scientist, and that fact was accepted; whatever opposition had existed, or that she had feared, was never again manifested.

To Rosalind, this was of enormous importance. How much she wanted his approval can be calculated by her lasting, sensitive belief that he, in particular, thought her graduate work at Cambridge a sign of failure. This may have been no more than a reflection of her own judgment. Her confidence had been shaken, temporarily, as it happened; but surely some of the passion with which she threw herself into her work at CURA can be accounted for by her need to prove her capacities once and for all, to herself as well as to all onlookers. In this she succeeded. As far as Rosalind was concerned, her road ahead was clear at last.

THREE

Paris

When the war was over, Rosalind was twenty-five, and somewhat further along the road toward a successful career than most scientists of her age. For most who try it, genuine success in science does not arrive rapidly, but cumulatively, in a slow progression, step by step. Science has had its *wunderkinder*, but they are not numerous; the nature of the subject works against them. Raw creativity may manifest itself in young artists long before they have mastered the techniques of their art, and often it is interesting, and sometimes it is valuable, too; but in any art there is a fairly wide latitude in which technical error or naïveté does not necessarily invalidate the work produced. Cimabue did not understand perspective, nor did Grandma Moses, but from looking at their paintings many people have derived varying degrees of pleasure. There is no real parallel to this in science, which abhors error, and which concerns matters in which naïveté cannot be distinguished from ignorance. If valid subjective perceptions may coexist with ignorance in a work of art without having their validity destroyed, valid objective perceptions which fly in the face of the facts are much harder to come by, or even to conceive of. A scientist, quite simply, has a lot to learn before originality can usefully assert itself, and even then, the originality may take a good deal of demonstrating, for a good idea does not stand independent of its proof.

In view of this, it is odd that a persistent legend exists which claims that all good scientists do their best work before they are thirty. It has been known to happen; it is rare. It happened more often a century ago than it does now, it has always happened in those areas of science which are most abstract—mathematics, for example—far more frequently than in those which are experimental. It is a foolish legend, and not quite an innocent one in its effects. There are cases known of young scientists who believed in it too firmly, despaired too quickly, and chose to hurry on the natural slow progression in inadvisable ways. A much more accurate statement would be that scientists tend to show their *potential* early, though when they fulfill it is at least partly a matter of luck. In Rosalind's case, the early potential was evident, and at an age when the average young scientist is barely emerging from graduate school, she had confirmed it with a substantial body of work.

She might have taken this as indication of where her future lay. Instead she was restless, even bored. That she ached for change is not surprising; by the end of the war, a great many people in England suffered from the sensation of having been cooped up in a narrow space for a long time, in circumstances which were sometimes dreary, almost always comfortless, and often exhausting. Apart from her work, Rosalind's life was unsettled. She had moved from her parents' house, first to lodgings more convenient to her job, then to the house in Putney that belonged to her uncle and aunt but which was occupied during the war by Rosalind's cousin Irene and a varying band of young women; at the end of the war, with Irene married and the house reoccupied by its owners, Rosalind went back to her family in Pembridge Place. She had had enough independence to dislike sacrificing any of it. Some of her restlessness was clearly a search for ways of asserting it, otherwise it is hard to understand why she considered even briefly, and never seriously, taking work as a games mistress in a girls' school in North Africa. This offer was a joke to her, but it was also not quite a joke. She had, after all, put herself in the way of such offers. She turned down the post, and stayed on at CURA, but not quite happily.

In the spring of 1946 she wrote to Adrienne Weill, who was back in Paris, in a tone that sounds faintly desperate under the lightness: "In spite of persistent efforts to move, I am still at CURA, which is still in its usual state of crisis. I am free to leave as soon as I can find another job. If ever you hear of anybody anxious for the services of a physical chemist who knows very little physical chemistry, but quite a lot about the holes in coal, please let me know." The letter had consequences. One of Adrienne's close friends was Marcel Mathieu, a distinguished scientist who held a directing position in the French government agency which supported, sponsored, and controlled the greatest part of the scientific research carried on in France. Mathieu was a remarkable man who had emerged from his wartime activities in the underground with a hero's reputation; on the postwar scene of French politics, he played an important role on the left; as both a scientist and a political man he was well known and much respected in important circles both inside and outside France. When he came to London in the autumn of 1946 to attend a conference on carbon research, Adrienne saw to it that he and Rosalind were introduced.

They liked each other very much. Adrienne's lessons in French bore fruit, for by then Rosalind had a fair, if not yet perfect, command of the language, quite enough to assure Mathieu that she could easily get on working in France. Beyond that, Mathieu was charmed, and when Rosalind read a paper at the conference, he was impressed as well. By the time he returned to Paris, he was Rosalind's friend, and this friendship endured. For the rest of his life, he retained a rather paternal affection for her, and for as long as Rosalind lived, she returned it with a respectful affection of her own.

He saw no reason why Rosalind should not work in France.

In February 1947, Rosalind went to Paris with an appointment as a *chercheur* in the Laboratoire Central des Services
Chimiques de l'Etat.

It came at the right time for her, of course. She went
abroad triumphantly. But that the time Rosalind spent in
Paris—from the beginning of 1947 to the end of 1950—was
probably the happiest period in her life is accountable for in
other ways than this. And it was not entirely attributable to
the fact that there was a good deal to be said, in any case,
simply for being young and in Paris at any time between the
end of the war and the mid-1950s. For Rosalind, all the
factors were favorable.

She found the atmosphere in general very congenial. Indeed, it was a good one for a young woman of intellectual
attainments with independence on her mind. France is not
England, and nowhere do they differ more than in their attitudes toward intellectuals, toward women, and toward those
who happen to be both. French culture is tenderer toward
intellectuals in general than any Anglo-Saxon one is. It incorporates a great respect for learning—too much respect, in
the view of some foreign critics, thinking no doubt of the
tiresome extremes that French pedantry can reach—and a
corresponding respect for intellectual workers, in which category any scientist naturally falls. This respect is less than
Germanic reverence, but it is friendlier by far than Anglo-
American mistrust.

French society at nearly all levels is interested in intellectuals, as well as in film stars and sports champions, and its
intellectuals respond by taking a good deal of interest in
society. They involve themselves in society's business, including politics, in a way which does not really occur to Oxford
dons or to Harvard professors; but then, neither the dons nor
the professors have quite the same sort of involvement available to them. There is no English or American equivalent to
Jean-Paul Sartre, for example, nor is one imaginable. There
is no English or American equivalent to André Malraux. No
American president or British prime minister has had among
his qualifications, as Pompidou did, a doctorate in literature,

and neither would have found one a political asset: Woodrow
Wilson never successfully overcame the "schoolmaster" image
attached to the learned man in American politics.

But in France it is different, and it is still more different
when it comes to women. For an English or American comparison to Simone de Beauvoir, no one is available. Though
France retained curiously repressive laws on its books for a
very long time which aimed at keeping women firmly under
the supervision of father or husband—and, for that matter,
has not repealed them all—France is a place where, as Mavis
Gallant says, "women on the whole have a better time of it
than in English-speaking countries." [1] Legal status is not
everything; it is the climate in which one lives that counts.

The climate in France, for women, is fairly sunny. "French-
men," Mavis Gallant writes, "do not seem to resent women
or be afraid of them, they are not bored by feminine company
(all-male clubs or outings are rare and considered ridiculous),
the war of the sexes scarcely exists. Equal pay for equal work
is the law of the country, and women often hold more important jobs than do women in America. A woman's intelligence is respected, her professional status accepted, and as
to her personal life, the French are notorious for an indifference to others that is also a form of minding one's business." [2] All this is true. It is what makes the climate sunny
most of the time.

It comes as a revelation to women who have been educated
in an English-speaking country, and who, as educated women,
have usually been confronted by unhappy choices: to be
meek and self-effacing on the one hand, or aggressively self-
defending on the other. Certainly it came as a revelation to
Rosalind, whose experience of the conventions had nowhere
included the customary assumption that respect for intelligence or acceptance of professional status were wholly detached from sex. It was what she herself believed, and was
willing to fight for, but she had not quite imagined a situation in which the fight was unnecessary. Many English and
American intellectuals have chosen to live in France because
they have found that certain undefinable tensions which

plague them at home vanish in French air, and for women intellectuals, whose tensions are greater, the sheer comfortableness of working in Paris can be nearly overwhelming. Rosalind was not overwhelmed, but she was responsive. Her contentment in Paris proves it.

In Paris, Rosalind formed some deep connections, friendships more enduring than any she had formed at CURA, for example, or was later to form at Birkbeck. She began, of course, with one friend in Adrienne Weill—who promptly proved her friendship by finding Rosalind a place to live in a city where accommodations were, at the time, virtually unobtainable. A bed-sitting room was located for her in the apartment of a professor's widow; a chief item of decoration was the professor's death-mask, executed in plaster, and the bookshelves were crowned with statuary so large that in some cases pieces had had to be curiously amputated in order to fit under the ceiling. It was far from luxurious, but it had the merit of being inexpensive, which mattered. Rather to Rosalind's delight, she was forced to live on her modest pay, as currency restrictions forbade the exporting of funds from England, and her pay did not run to luxuries. But she liked her room, for all its oddity; she got on reasonably well with the professor's widow, who had taken her in only at Adrienne's request, and who never was entirely reconciled to having a lodger; and she got on excellently well with the professor's widow's maid, who gave her cooking lessons, and often supplied her with a surreptitious meal.

Beyond Adrienne, she made friends at the laboratory, which was a happy place. It was located on the quai Henry IV, in an old building never designed for the purposes for which it was put to use, but which, perhaps because of this, had—for a laboratory—quite unusual charm. The rooms were large and lofty, a little bleak in winter when the light that came in through the many windows was gray and cold, but rather romantic in summer when the slightly dusty beams of sunlight glittered on the laboratory apparatus.

The physical surroundings encouraged informality, but the people who worked there would probably have insisted upon it anyway. They were enormously interested in what they were doing, but they were also very interested in each other. They had the European interest in personality which so often embarrasses those who are not European. They consequently quarreled, gossiped, indulged in tremendous intellectual debates, fell in and out of love, exhibited a frank range of human passions, and much candor, and on the whole enjoyed each other thoroughly. They made up a society outside of working hours as well as inside, lunching together at whichever inexpensive neighborhood *bistro* was in current favor, exchanging dinner invitations, picnicking on weekends in the pretty countryside around Paris, and often going off in a group for a holiday of skiing, mountain climbing, or camping. Into all of this Rosalind fitted easily, and with great pleasure.

What made it easy was that she was instantly recognized as a personality, a passionate nature, a temperament. Those who were working with her read her reserve, correctly, as a form of shyness disciplined by a proper English upbringing, but they did not confuse it with coldness or a phlegmatic emotional organization. They did not find her a little English mouse, and were pleased that she was not. All was never sweetness and light at the laboratory on the quai Henry IV, too many strong personalities were gathered there to make this likely; but when Rosalind was at the center of a storm, as sometimes she was, those around her took the storminess as proof of sensibility and susceptibility, both qualities worthy of admiration.

But it was not all storm either. With Jacques Méring, with whom Rosalind worked closely for most of her time in Paris, she had a warm and understanding relationship. She was, in a sense, Méring's pupil to begin with, learning from him techniques in X-ray diffraction, a science new to her; subsequently, they collaborated extensively, and the results of what she learned from him can be seen in her series of papers on graphitizing and nongraphitizing carbons.[3] This was the beginning of Rosalind's work in crystallography, and it bears Méring's mark, in that she entered a difficult field at a difficult point. His special interests lay at that time in such amorphous substances as graphite, which demand not only remarkable technique simply in the handling, but which require a strong theoretical mind for the interpretation of experimental results. Both the technique and the theory he passed on to Rosalind, who could not have gotten them from a better source. She used them quite brilliantly in working with biological substances, some of which are also amorphous, and all of which pose exasperating problems in X-ray diffraction. What this introduction to crystallography implied, however, was an untraditional approach, through materials which barely fell within the definition of crystalline, and that left her unexposed to what might be called a classical education in the subject.

Rosalind had great capacities to pick up what she needed as she went along, and to put all her experience to good use. These were among her strengths as a scientist. In crystallography, it was also a strength to her that she developed in Paris unusual abilities for working at the upper end of a scale of difficulty. But occasionally she lacked what a more fundamental training might have provided, for sometimes pure crystallographic information carried fewer implications to her than it might otherwise have done. Because she was aware of this, she never thought of herself as a crystallographer, only as someone who used the methods.

Probably her closest friends in Paris were Vittorio Luzzati, who also worked in the laboratory on the quai Henry IV, and his wife Denise, who was then a medical student. In the laboratory, Vittorio was Rosalind's ally, her *copain,* whom she teased and wrangled with, turned to for sympathy when she needed it, and competed with for mastery of the French language. Vittorio had in this competition the advantage of a French wife, and Rosalind of a superb ear for the spoken word. For music she had no ear at all; the best a school report could say of her musical accomplishments was a glum, "improved, she now sometimes sings in tune;" but her French rapidly passed from good to excellent, and before she left Paris she was credited by her colleagues with speaking the best French any of them had ever heard in a foreign mouth.

Rosalind and Vittorio liked each other, enjoyed each other, respected each other, and Denise kept them both in order when their temperamental flights grew too extreme. Until Rosalind's death they remained very close.

And there were others. The circumstances encouraged a capacity in Rosalind for making friends that sometimes, in other situations, her natural reserve blanketed. After Rosalind had left Paris, it was this to which she looked back with longing. The circumstances were never duplicated elsewhere. Perhaps they had something to do with time, with youth, with a short moment in history that could in no event ever be repeated, that might not have endured unchanged even if she had stayed where she was. But certainly this period had done more for her than simply teach her a new branch of science, advance her career by a giant step, and confirm her in happy independence. It also provided her with a confidence in herself that was serener and more personal than any she had previously known. This is a concomitant of growing up, and what Rosalind had accomplished in Paris, in addition to a great deal of good work, was the last of her maturing. She was always young for her years, but henceforward it was a purely charming trait. The uncertainty was gone, and it never came back.

FOUR

The Problem

The technique Rosalind had learned and worked with in Paris —X-ray diffraction, X-ray crystallography, chemical crystallography: all three names are applied to it interchangeably— occupies what might be considered a corner of science, and because it is a field possibly distinguished among others by its modesty, it rarely presents itself as more. From the beginning, crystallographers habitually made rather small claims for what they were doing. They were to all appearances surprised and delighted when it turned out that what they were up to was of considerable importance; this diffidence has to some degree lingered, even though the importance has become steadily more and more unarguable. Such modesty is unusual among the sciences, which do not often embrace understatements of their own potential and which are also, for the most part, practiced by people with no great gift for understatement either. Where this odd and pretty habit came from it is difficult to say. Nevertheless it exists, and X-ray crystallographers, now in the third or fourth generation of their kind, are consequently just a little different from scientists in other fields: friendlier, for one thing, and less assertive for another.

When Rosalind took up X-ray diffraction work, the subject was rather more than thirty years old and had reached the point of rapidly expanding horizons. The early history was simple. It begins in Munich in 1912, where Friedrich Knipping,[1] working under the direction of von Laue, demonstrated that when a beam of X-rays is passed through a crystal, the beam is scattered in such a way as to record upon film a visible pattern—the X-ray diffraction pattern, which varies from substance to substance, but which is always identical for identical substances.* If it appears that not much more was provided by this information than a kind of fingerprint confirming identification, then something else was quickly added. Very shortly after the Munich work was published, W. L. Bragg read a paper before the Cambridge Philosophical Society in which he described the manner in which the atoms must be arranged in a crystal of zinc blende in order to give rise to the specific diffraction pattern observed and recorded for zinc blende.[2] This may be taken as the moment at which X-ray crystallography was born. It had at that point a limited future. W. L. Bragg's method for accounting for the pattern was trial-and-error. He hypothesized models until he found one in which the hypothetical arrangement of the atoms precisely fitted the diffraction effect and, in so doing, created a very good method still sometimes used; but it is a laborious one and efficient only when applied to substances the structure of which is either simple or regular in its symmetry. But not long after—in 1915, in fact—W. H. Bragg, the father of W. L., produced the next essential step, after which X-ray crystallography could, indeed, be named a new field of science, and, as it happens, potentially a very powerful one, too.

The simple explanation is that X-ray crystallography is a form of microscopy, except that the microscope is lacking. It is replaced by what may loosely be called a type of artificial vision, the virtue of which is that it can perceive—though not precisely "see"—things of an extreme minuteness, in this case,

---

* This is a not-quite-accurate statement, in that many substances are capable of crystallizing in more than one form, and each different crystal form provides a somewhat different diffraction pattern. But all crystal forms of the same substance will show *related* patterns.

the separate atoms which make up a given molecule. The method of doing this is delightfully simple in concept, and often exasperatingly difficult in practice. The beam of X-rays passed through a crystal of the substance under examination produces a pattern upon film; it cannot produce anything more than this because there are, up to the present time, no materials from which lenses can be made that will focus X-rays. But the function and behavior of the missing, and unobtainable, lens can be provided—to be exact, imitated—mathematically.

This is precisely what W. H. Bragg proposed, using for the purpose a mathematical process called the Fourier transform. Therefore, from spotted films which look to the uninstructed like nothing very much, and after the application of extensive and intensive mathematical analysis, molecules can be made to yield up the secrets of their inner and invisible anatomy, and three-dimensional representations can be made (usually in the form of plastic models or maps) which locate the position of every atom which goes into the making of the molecule in question, exactly as if this subminute organization has been literally magnified to visible range.

This is X-ray crystallography, chemical crystallography, whichever one cares to call it—an elegant concept, and quite evidently a powerful one, if what one has in mind is the detailed discerning of what is otherwise undiscernible. With such potential in hand, it is a little surprising that crystallography launched itself unpretentiously and proceeded to advance at such a casual pace. The early publications in the subject reflect the casualness; they have a kind of engaging, informal charm, as if they represent amiable communications exchanged among a small group of friends, fellow hobbyists who went about their business seriously enough, but for the pleasure of it, at leisure, and without much concern for where it might all be going. What is noticeably missing are those grand predictions of the wonders the future will see accomplished which are common to scientists, which may explain why the wonders of crystallography, as they have come along, have often been met with that sort of pleasure which is the nicer because it embraces a slight element of surprise.

The wonders have come about. Crystallography began quietly, chiefly as an adjunct of metallurgy and mineralogy; in the late 1920s, it expanded into the study of the structure of organic compounds; it grew steadily more sophisticated; but that it would come to be of major importance in the elucidation of the nature of biological substances was not quite predictable. This has happened. And it is because the horizons of crystallography were rapidly expanding to the point of including biological substances that Rosalind left Paris and went to King's College.

It is reasonable to ask why she was interested. She was not in any sense a biologist. She was by training a physical chemist who had achieved an exceptional command of her field, she was by experience a highly specialized researcher working exclusively with carbons, and the connection between what she had done and what she was about to begin doing might seem obscure. But one consequence of learning anything new is that one usually learns more than one anticipated.

Problems in science converge, especially at fundamental levels. The importance of molecular structure—what crystallography aims at delineating and establishing—was early recognized by those who dealt in metals and minerals, where it could often be plainly seen that the architecture of a molecule had much to do with its behavior and function; by the 1930s, biologists had begun to suspect that the same might be true of biological molecules. This conviction grew, though there was no proof of its truth, simply because there were no significant biological substances whose molecular structures were known. X-ray diffraction methods were plainly one of the most promising approaches to an understanding of biological substances and, in the long run, have proved one of the most profitable. But nothing was then so simple, for X-ray diffraction methods encounter increasing difficulties as the size of the molecule increases, and most molecules of biological importance are very large ones, producing diffraction patterns of be-

wildering complexity which for a long time lay outside the
range of possible mathematical analysis (considering, that is,
that human lives are of finite duration, and the amount of
mathematical labor which can be crammed into any one of
them naturally limited °). But, in the late 1940s, an acceler-
ated interest in the structure of biological molecules was
matched by some improvements in crystallographic methods
accompanied by daring, which is exactly how new frontiers, in
science or elsewhere, are opened.

That all of this communicated itself to Rosalind is not surprising. And that she wanted to do something about it is, in
view of her nature, less surprising still. She wanted this very
much, for nothing less than a new and demanding challenge
could easily have tempted her away from Paris. If she had
given first importance to her personal life, she might never have
left, for she was still genuinely contented there, and certainly
there was at that point nothing to force her away.

It is true that her position was not entirely secure. As a
chercheur of the Centre National de la Recherche Scientifique
she was that anomaly in a French institution, a foreigner; but
any threat that this implied lay in the future and, in any case,
might never have become active. Her foreignness may also
have made her somewhat immobile, but a lack of prospects of
rising in terms of the organization would not have troubled
Rosalind in the least—her ambitions were never for status or
power, and on later occasions she voluntarily avoided the
sort of promotion that gratified only with power or status and
otherwise distracted from what she had in mind to do. A sort
of sideways immobility, in the sense that she might have found
it difficult to move from one CNRS laboratory to another,
could have affected her more; but there is no indication that

---

° Max Perutz began to work on the structure of hemoglobin in
1937, and achieved it twenty-five years later. This was a prophetic
piece of work as well as a triumphant one. That a protein structure
was susceptible of solution within the span of one man's life cannot
have been in the least evident in 1937; that it was accomplished
was both the justification of a vision, and a measure of the rate at
which crystallographic techniques were advancing.

she made such an effort, and it is not likely that she did, for in all probability there was no other French laboratory which particularly attracted her.

What was attracting her was the possibility of doing X-ray diffraction work on biological substances, and this was a vast attraction, indeed. Because of it she was willing to leave a place in which she was happy and contented for one where she had no great desire to be. She may have felt some conscious apprehensions concerning the chill of English intellectual life compared with what surrounded her in Paris, but what she mentioned was the "boringness" of England where—with currency restrictions on foreign travel as strict as they then were —one could so easily feel trapped. And surely she hated to leave her friends. Those who are reserved by nature, and who rarely make friends quickly, or lightly, have a natural reluctance to say good-by, if only because new relationships will not quickly, or lightly, replace the ones that are left behind. If Rosalind was not friendless in England, and she was not, she was all the same equipped there with few acquaintances of the sort she had in Paris—scientists like herself, absorbed in the same problems, involved in the same interests. She was too familiar with English life to assume that she would easily find in a new situation people with the outlook that so delighted her in France, and that had, in a sense, Europeanized her.

No doubt she hoped. No doubt she also suppressed her misgivings. The move might have, from a personal point of view, little to recommend it, but Rosalind was also eager to try her hand at the new phase of crystallographic research, the biological phase, and the immediate opportunity to do that lay in London, not elsewhere. Her decision was dominated, as decisions were very likely to be in Rosalind's case, not by impulse, but by where the work lay. On the other hand, it would be ridiculous to assume that she was prepared to seize an opportunity at any cost. But this is assumed, tacitly rather than overtly, by almost everyone who writes about her connection with King's College, in the simple repetition—which by now has become tiresomely fixed in the literature—of her pilgrimage there in order to become Maurice Wilkins's assistant. It seems to impress no one that Randall, who was responsible for importing her, denies that this was the capacity in which she was brought into the laboratory of which he was the director, nor has it helped very much that Wilkins, who also ought to know, affirms the flat denial. It may not, therefore, help very much to point out that from Rosalind's point of view such a move would have been absurd, but certainly this was the case.

More was involved in this than pride. Rosalind's was not a self-analytical nature, but there were things about herself which she knew because they had been objectively tested and proved, and because these things referred to her work, she was willing to be governed by them. The greater part of her work, and most of the best of it, had been done independently, not in close collaboration, not under direction. Not only was she happiest in independent work, she was more productive working so, and of this she was perfectly aware. She was also aware that she was insufficiently pliable and much too impatient to succeed very well as a categorical subordinate, and nothing would tempt her to try it. Given *liberté* and *égalité*, she entered wholeheartedly into the spirit of scientific *fraternité*; she had proved this to some degree at CURA, and had proved it quite fully in Paris; she was entirely happy with this state of affairs, and she had no intention of abandoning it either in principle or in practice.

If the best that King's College could have offered her was the chance to assist in someone else's research, she would simply have waited for a better opportunity, and with some confidence that one would be forthcoming. After all, pride does enter in. If Rosalind was free of false pride, she was also free of false modesty. It might be pointed out that professional workers in any field almost always are; it is the amateur who casts down his eyes, and murmurs of whatever he has done that it was nothing, really, just a fluke. But what distinguishes even gifted amateurs from professionals is that the amateur refuses, for some reason or another, to make the professional commitment, to accept not so much the importance of his own function as the importance of the work itself, and on the basis of this commitment to find the strength to abandon not only excuses, but modest deprecations. To Rosalind, who was in a very high degree professional, the work that she did was of the most unarguable importance, well worth giving her life to. Anyone who served it well was engaged in an activity which could not be deprecated, especially as to do so was to endanger the objective standards by which such service could be judged. This is not egotism; but anyone who feels this way is obliged to perform at the highest available level, and is also obliged to rate his own accomplishments sternly, valuing what deserves to be valued, and regretting what does not. Rosalind possessed gifts, and she knew that she did; she did much good work, and she knew how good it was. To become a research assistant at her stage of development as a scientist would have struck her as wasteful, and because wasteful, as quite absurd.

She did not consider it; but then, the question never arose. It is difficult to know what those who have characterized her in this role, beginning with Watson in *The Double Helix*, have had in mind. Certainly they have either been ignorant of, or have chosen to ignore, what work Rosalind had to her credit. The publications which resulted from her time at CURA were distinguished; those that emerged from her years in Paris were yet more so. She went to Paris to learn a new method, and certainly she learned it thoroughly, for all that neither her own approach nor the training she received were precisely along the lines of a standard education in crystallography. What she had from her work, first with Jacques Méring, and then on her own, was in many ways better. There is nothing like working with substances too amorphous to possess much discernible structure, which therefore yield up very little evidence in the way of clear diffraction patterns, for developing a sensitivity to what can only be called crystallographic *nuance*. This Rosalind never lacked, and this she was forever afterward putting to good use.

The handsomeness of her carbon work deserves emphasis; biologists tend to pass it over much too lightly. The carbons on which Rosalind worked were not among those which crystallize into neat geometries nicely reflected in reasonable diffraction patterns, but the sort which produce foggy and indeterminate evidence. Not many people were willing to try them; she mastered them. She was intensely productive. Between 1950 and 1956, she published eleven papers dealing with carbons, only one of which was coauthored, and all of them represent retrospective work, for by 1951 she had moved on to working with DNA. A few of these papers were minor, but several were of major importance, and a rate of productivity that works out at rather more than two papers a year is testimony not only to great industry, but to the possession of the scientific equivalent of a creative mind. These papers in themselves were sufficient to make a reputation had she never done anything else, and all of them were based upon research which Rosalind had accomplished before she reached the age of thirty. What they contain is beyond brief description; curiously, the technicalities of carbons are more difficult to understand and to communicate than those concerning much more complicated biological substances. In this case, to describe them would not contribute much to a grasp of what Rosalind later, and still more importantly, accomplished. But it ought to be said that this series of papers expresses a remarkable inventiveness in experimentation, and a capacity for essential insights into the interpretation of very elusive structures that is more remarkable still.[3]

None of this is beside the point. When Rosalind was confronted with DNA—an amorphous substance, difficult to handle experimentally, tiresomely recalcitrant from a crystallographer's viewpoint, requiring acute perceptiveness, if the scanty data it provided were to be interpreted at all—she was neither inexperienced, nor lacking in the needed arts and instincts. And it was DNA that she encountered when she went to King's College.

There were other problems available. When Randall offered Rosalind a Turner-Newall Research Fellowship, it was on the understanding that she would be put in charge of building up an X-ray diffraction unit within the laboratory, which at that time lacked one, and this is what she set about doing. She had not been brought to King's College to work upon DNA or any other specific problem; most sizable laboratories, and especially teaching ones, have a number of research projects going on simultaneously, and any university department of biophysics is certain to have several in hand at any moment to which the application of X-ray diffraction methods is desirable, or appropriate, or essential. The available problems were numerous, but of all the projects under way in the biophysics department of King's College in 1951, the one devoted to the investigation of DNA was not only the most important, but to any imaginative scientist, the most provocative and fascinating.

Rosalind could not resist it, and for good reason. What most people know of DNA is that it has something to do with heredity, a subject with built-in fascination; what fewer people have had reason to consider is the history of research into the mechanism of heredity, or the places where this research has led, although these are matters more fascinating still. Heredity has been on people's minds for a very long time, very probably since that lost day when the first understanding dawned of the nature of sexual reproduction; it has been rediscovered and restated as an operating fact in countless places; it has been a familiar mystery, but different from other real, but mysterious things—bolts of lightning, plague and famine, tremors of the earth—because it affects not some of us some of the time, but all of us all the time. It is, in fact, our intimate, inborn, inescapable, inexplicable destiny. Common experience confirmed its reality and was variously recorded: the sins of the fathers were to be visited upon the children by inheritance; Adam's fall was passed on in the moment of conception; the Habsburg lip was a matter of objective observation and could be taken as guarantee of the legitimacy of the heir. There is really no end to our long-standing acceptance of heredity as a fact, and no beginning either: the Bible is full of good advice, no doubt carefully compiled from long experience, about the planting of selected seeds and the management of pure-bred herds.

To put it simply, from the first translatable human records to the present, there is ample evidence that people have assumed that heredity works. But how? The answer to this is very far from obvious. It is interesting to reflect that if we are alive now, and perhaps are middle-aged, then we can assume confidently that what our great-grandparents knew about the workings of heredity was just about equal to, and not materially different from, what was known to the prophet Isaiah or to Alexander the Great. Not for centuries, but for thousands of years, ordinary observation suggested that some characteristics are inheritable, are literally "handed down" from parent to offspring, and ordinary observation also took note of oddities: that not all offspring strongly resemble either mother or father, that "throwbacks" to earlier ancestors are sometimes apparent. But observation really stopped there, and to the question of how such things can be, it offered no answer. This is not to say that explanations were never forthcoming. There have been thousands of them, random guesses, shots in the dark, of which perhaps the best was Aristotle's. Not surprisingly, he had the details wrong, but he came close to the truth in suggesting that what biological inheritance implies is the transmission, somehow, of what might be called a plan for development, passed on by some means or other from parent to offspring.

The first step toward solving the mystery of how heredity works took place in 1865. This is an arbitrary statement, but it is hard to argue convincingly for an earlier date. A century before Linnaeus laid down something of a substructure for systematic thinking about heredity by pointing out that on the basis of resemblances maintained throughout many generations, plants and animals can be classified into groups and families, but this in itself suggests nothing about how these persistently maintained resemblances can be accounted for in practice. In 1859, the publication of Darwin's *Origin of Species* provided a splendid opportunity for some new and insightful guesses, but the opportunity was lost, for in the storms of controversy that raged about the whole troublesome business of evolution, purely evolutionary factors in the family trees of organisms were emphasized almost to the exclusion of the interesting question of the ways in which, exactly, these adaptations might occur.

But in 1865, Gregor Mendel published a paper that became the ancestor of all the modern work on the mechanism of biological inheritance that was in the end to explain the mystery. Nowadays children learn about Mendel in high school, if not earlier, and most of us retain from some educational experience the vague and charming picture of a monk pottering gravely in the pea patch of his monastery's garden in order to produce the neat black-and-white diagrams in the textbook. That Mendel's discoveries went unnoticed until they were "rediscovered" in 1900 by Hugo de Vries is less well-known, though this extraordinary inconspicuousness of what was, indeed, great scientific work is understandable when one realizes that Mendel published in the *Journal of the Brno Society of Natural Science,* a publication which can never have achieved a wide circulation.

But what did Mendel do? He grew peas, just as we all remember, but he chose strains to grow which exhibited certain particular characteristics and then cross-bred them with strains having opposed characteristics—wrinkled seeds, say, as opposed to smooth ones. From the observed results in the progeny, he postulated a wholly new theory which said, in essence, that where an organism reproduces itself sexually, characteristics are inherited according to a determinable pattern involving "units of inheritance" located in the reproductive cells of the parents and that, consequently, the distribution of characteristics among offspring can often be quite accurately predicted and described by a simple mathematical formula.

It sounds so extraordinarily simple, no doubt because nowadays we have come to take it all for granted, as common knowledge. But nevertheless it represents the first step in man's long recorded history toward the solution of his oldest puzzle. What is not to be overlooked is the idea that inheritance arises out of finite, potentially locatable and describable "units" possessed by parents and passed on to progeny.

It might be said that this idea bridges the gap between the unimaginable and the science of genetics, for the unit of inheritance is—though at this point only theoretically—the gene.

This is not to say that a shaft of light illuminated all the darkness. It never does, not even in science, which is, after all, one of the most cumulative processes man has devised. In doggedly practical terms, Mendel's work illuminated nothing until 1900, for if the *Journal of the Brno Society of Natural Science* found its way to library shelves far from its native Moravia, it remained on the shelves, and Mendel died in 1884 without knowing that he was to possess a place in history. But other discoveries were being made which, when brought together with those of Mendel at the appropriate moment, created quite a source of light.

Cells were receiving a good deal of attention in the nineteenth century, for although they had been described as early as 1665 by Robert Hooke, they did not reveal much of their nature until improved microscopes permitted visualization of what was previously unseen. Hooke's notion of the cell was of a rather dead thing, but in the 150 years that followed his observations, the cell was shown to be alive, busy, capable of growth, motion, and multiplication; one-celled animals and the complex organization of other animals made up of many cells were noticed; and, in 1839, Theodor Schwann suggested that plants and animals were "aggregates of cells which are arranged according to definite laws," though what these laws might be remained undefined. By the middle of the nineteenth century, the chief components of cells—nucleus and cytoplasm —had been discovered and Rudolf Virchow had proposed that all organisms are not only composed of cells, but that these can derive only from pre-existing cells, a notion widely accepted from the moment when it was announced and today taken as a maxim of obvious truth. From this arises another obvious truth: that all cells existing in an adult organism can be traced back through a series of divisions to a beginning in the fertilized egg. The history of living organisms, if not their method of passing on their inheritance, becomes that much clearer. Not long after this was perceived, the presence was observed in the nucleus, when a cell was undergoing division, of microscopically visible, numerable, chemically stainable, threadlike bodies: chomosomes.

It was against this background of additional information that Hugo de Vries, and—quite independently—Carl Correns, "rediscovered" Mendel's work. Against this background the suggestion could be made, and quickly was made, that Mendel's "unit of inheritance," by this time rechristened the gene, had its identifiable home in the chomosome. It should be noted that the gene remained theoretical, a concept; but chromosomes were, happily, visible and manipulable, and were to that extent suitable subjects for experiment as well as for observation. Upon this the modern science of genetics was launched.

There was good evidence that the genes resided in the chromosomes. Consider, for instance, that every cell which composes an organism has its due and fixed number of chromosomes, varying from species to species.* There is no exception to this chromosome-population count except in the sex cells, the sperm and the egg, which possess only half the number found elsewhere. What can more reasonably be assumed than that the *diploid* cells, arising as the cell theory dictates from the pre-existing fertilized egg, must have acquired their double complement of chromosomes by the simple addition of the two *haploid* sets present in egg and sperm? Nothing else fits the arithmetic.

Classical genetics did better than this. T. H. Morgan worked for years on the genetics of the fly *Drosophila* and eventually demonstrated that the behavior of *Drosophila's* chromosomes not only so closely paralleled the requirements laid down by Mendel's theory that the association between theoretical gene and visible chromosome might be firmly made, but that there was direct evidence as well to indicate that an identification could be made between a specific gene and a particular chromosome, even to the extent that the various genetic characteristics of the fly could be mapped to precise locations on the chromosome. And as the total number of genetic factors specified was vastly larger than the number of chromosomes involved in any division, then the genes must indeed be contained within the chromosomes.

---

* Mendel's peas, for instance, though he was unaware of it, possessed 14 chromosomes—tomatoes have 24, mice 40, human beings 46, horses 64, and some protozoa 1,600; and to every other species on the face of the earth that reproduces sexually an appropriate number may be fixed.

But how? From a chemical point of view, chromosomes are made up of a very limited number of constituents. One is protein and the other we now know as nucleic acid. In 1868 it was described by Friedrich Miescher under the name of "nuclein" as a phosphorus-rich acid substance; twenty years later, A. Kossel took the matter further, and identified it as a substance composed of four nitrogenous bases, a 5-carbon sugar, and phosphoric acid. He also changed its name. If chromosomes are composed of only protein and nucleic acid, then one or the other must act in the role of a transmitter of highly complex genetic information—but the question is, or was, which? Proteins are complex, made up of subunits called amino acids linked together in chains; each link of the chain may be any of twenty amino acids. What would seem more likely than that the complex genetic information would find suitable storage in this large protein molecule with its rich vocabulary of interchangeable units? The concept is an attractive one, very nearly overwhelmingly attractive, and it proved by no means easy to dismiss. In comparison, nucleic acid seemed ill-fitted for such an elaborate function; but it must not be thought that because the odds were unpromising, nucleic acid went uninvestigated.* P. A. Levene and his associates, in particular, did an exhaustive chemical study from which emerged, to begin with, the identification of two separate nucleic acids: RNA, or ribonucleic acid, and DNA, or deoxyribonucleic acid, which differed in their nucleotides, or smaller subsections.

Something else emerged, which was an incorrect theory. It did not concern the chemical constituents of the molecules in question, but their structure. This hypothesis concerning the structure was not based upon the sort of evidence available from such methods as X-ray diffraction, because there was no possibility in 1926 of using such methods upon any molecule as large as those of the nucleic acids. It was purely theoreti-

---

* A small idea of the amount of work that went into this research is provided by a review article published in 1939 that cited 116 papers reporting work done upon one or the other of the nucleic acids, and nearly all the papers cited were of recent date.

cal and, for all of that, it was partly correct. It was also ultimately misleading, and the part that was wrong had the unfortunate effect of dogging most thinking about nucleic acids
for more than twenty years.

Levene and Simms were correct in suspecting that RNA
was a molecule possessing a sort of spine composed of a link-
age of phosphate to sugar to phosphate, with other components
—the purine and pyrimidine bases—"sticking out." They were
incorrect in suggesting that the whole molecule was a short
one, made up of only four nucleotides, or subsections. What
was suggested by this was that nucleic acids might be made
up of molecules composed of flat rings of atoms, perhaps in
some way stacked, and what this in turn implies is the discouraging unlikelihood that any such simple rings could contain, and be able to pass on, genetic information. It may be
said that, with the emergence of the tetranucleotide theory,
it began to appear that the more chemists learned about
nucleic acids, the less reasonable it appeared that they could
have much to do with the mechanism of heredity—not, at least,
as long as the nucleic acid molecule remained fixed in concept
as a mere four units long, and monotonously repetitive.

So things stood for quite some time. As late as 1935, a
thoroughly up-to-date and perceptive paper by Dorothy Wrinch
on the molecular structure of chromosomes could simultaneously anticipate, in a rather prophetic way, a science of molecular genetics still twenty years in the future, and incorporate
a wrong guess because the tetranucleotide theory allowed no
other. Although she realized with great insight that the linear
nature of genetic information demanded a linear molecule for
its expression, she was required when faced with the choice
between protein and nucleic acid to opt for protein, surmising
that the nucleic acids might play no more of a part than that
of a kind of glue that held the chromosome together.

An enormous amount of very complex work in the biochemistry of genetics went on between, say, 1925 and 1944, all
of which is here elided; [4] for present purposes, what matters is
the experimental evidence published in 1944 by O. T. Avery,
C. M. MacLeod, and M. J. McCarty which indicated very

clearly that, where bacteria were concerned, bacterial DNA appeared to be the carrier of bacterial heredity—a conclusion which amazed, even to the extent of surprising the authors of the paper. It was too amazing to be readily accepted, though a good deal of confirmation came in steadily, all of it incompatible with the notion that DNA was a large invariable polymer made up of those four monotonously repeating units.

In 1950 the tetranucleotide theory was finally put to rest. Erwin Chargaff, using methods of paper chromatography which made possible very exact measurements of the components of DNA, demonstrated that the four bases present in DNA—adenine, guanine, cytosine, and thymine—were present in varying proportions according to the biological source from which the DNA was taken: not uniformly, not monotonously. And, to quote Gunther Stent, "The way was now clear to formulate a theory of how DNA can act as the carrier of genetic information. . . . It seems impossible today to establish who was actually responsible for originating these notions. The theory suddenly seemed to be in the air after 1950 and had come to be embraced as a dogmatic belief by many molecular geneticists by 1952. The key proposition of this theory is that if the DNA molecule contains genetic information, then that information cannot be carried in any way other than as the *specific sequence of the four nucleotide bases* along the polynucleotide chain." [5]

This, then, may be taken as the general state of affairs with respect to DNA when Rosalind went to King's College. The problem was irresistibly attractive. The oldest of questions—how heredity works—might be on the verge of finding an answer, and that answer might lie in discovering the structure of the molecule. It was, of course, chancy. On the one hand, delineating the structure through X-ray diffraction methods might well be impossible. Previous attempts had not gotten very far beyond indicating that DNA was a poor subject for X-ray photography, producing very little diffraction data, very possibly too little to allow any significant interpretation. On the other hand, there was no guarantee that the structure, even if accessible, would provide a clear answer, for the answer might lie still elsewhere. But clearly the attempt had to be made.

Rosalind was determined to make it.

FIVE

"One Cannot Explain
These Clashes
of Personality"[1]

Rosalind went to King's College full of high hopes. What occurred there between the beginning of 1951 and the spring of
1953 constituted a bitter disappointment to her, a fact which
ought to be obvious, but which has not been mentioned by
any of the people involved in the events to whom I have talked
and which was certainly—though naturally, one supposes—
overlooked in *The Double Helix*. From the viewpoint of those
connected with King's, the disappointment was perhaps too
general, and equally too personal, to be considered in anything
but their own terms; and, as far as Watson is concerned, *The
Double Helix* is his record of a triumph which does not waste
much time on sympathy.

Rosalind's disappointment, however, was a reality. It began
almost as soon as she arrived. And it was not quite a trivial
disappointment either, to be regarded only in terms of an in-

dividual for whom things worked out rather less well than they might have done. Even with the advantages of hindsight, it is impossible to guess at any point the extent to which the course of history has turned upon the ability of two people to tolerate each other. All that can be said here—though it deserves saying—is that it is very possible that the history of molecular biology might be rather different from what it is today if Rosalind and Maurice Wilkins had not hated one another at sight.

This is not really an exaggeration. There is no record of a relationship which began promisingly and then degenerated into hostility. Only too evidently the antipathy was instant and mutual. All instant antipathies are a puzzle, and this one is no different; the explanation which will truly explain it will forever be lacking. There is no reasonable argument available to either side. Two people who ought, on the face of it, to have had enough in common to be able to get along—and the situation did not require much more than merely getting along —could not quite manage it. Randall's comment is a fair one; as he rightly said, "One cannot explain these clashes of personality," and there is no use in trying.

It is not beside the point to observe that such clashes, particularly when they become so extreme as to be damaging to all parties, rarely take place in a vacuum. When two people not destined by nature to become close friends meet, and fail to become friends, there still remains available a wide range of cool and indifferent behavior, on the whole neutral, perhaps glazed with politeness, which usually serves to bridge the gap. In more than one place can tepid colleagues be found, who cannot be described as feeling much enthusiasm for one another, but who nevertheless contrive a *modus vivendi*. Rosalind and Wilkins were not only alienated, but hostile, and sometimes actively so, and this is sufficiently unusual to be unaccountable, unless one assumes that something in the surrounding circumstances was extraordinarily unpropitious.

This exactly may well have been the case. And if this was so, then everyone was to blame for what happened and no one was to blame. The question remains: why did it happen?

One may as well begin by observing that the circumstances were much less favorable to Rosalind than to Wilkins. For one thing, he had been at King's for five years; he is still there, which allows one legitimately to assume that he continues to find his situation a pleasant one, and that the organization takes pride in him. In 1951, he was simply an old hand while Rosalind was a new one. But in 1951, he also had the additional advantage of being a man. In those days, King's College, as an institution, was not distinguished for the welcome that it offered to women.

It has changed since; things are not what they used to be; but what they used to be had its unattractive side, and the explanations usually provided to account for this explain a good deal more than they excuse. They point out that King's College was founded as a theological school; that as the Church of England does not admit women to its ministry, it was naturally a male institution; that as a Church of England foundation, it took its early inspiration from older, male, theological foundations at Oxford and Cambridge and therefore cultivated traditions which did not appreciably alter when the college ceased to be theological or began to admit women. This explanation can never have seemed more satisfactory to the women who found themselves at King's than it has seemed to me on the several occasions when it has been offered. The best that can be said for it is that it is ingenuous, not unlike apologizing for one's insulting behavior on the grounds that one is habitually rude.

It does not make as much difference as perhaps it ought that the lack of welcome given to women under King's fixed traditions mostly expressed itself in minor ways. Great injustices have the merit of earning sympathy, and often some redress, for those who suffer them; but those who are subjected to small, daily annoyances may well be considered, if that is how one chooses to look at it, as rather irritating and unsporting if they complain. No doubt it was pleasant that in 1951 women were allowed to earn degrees at King's and to work on its staff and in its laboratories, but it cannot have been entirely agreeable to the women who exercised these privileges to be kept in a kind of purdah while they did it. Quite simply, it was a great advantage in those days to be a man if one was connected with King's College.

Rosalind was not a man. She was unused to purdah and often it offended her. Part of the surrounding circumstances was that, from the start, she was dealt with at King's less as a scientist than as a woman, hence inferior. The inferiority has been deduced, but there is evidence which implies it. It is a minor thing, but perhaps not so very minor, that in those days the male staff at King's lunched in a large, comfortable, rather clubby dining room, though the female staff—of any age or degree of distinction whatever—lunched either in the students' hall, or off the premises. This odd exclusiveness no longer prevails. A man who has been connected with King's for many years remarked that he regrets the change, that the men's-club atmosphere had been pleasant and relaxing, that he wishes it might be restored. "Now you are a sensible woman," he remarked to a visitor, flatteringly, "Would that sort of thing bother you?" The only honest answer seems to be yes, the more sensible the woman, the more she was bound to be bothered. The nonsense of it, if nothing else, would offend.

There is also more than nonsense to this sort of arrangement. Quite apart from the selfishness and insensitivity which can effortlessly be read into this way of doing things—pleasure and relaxation for part of a society, an indifference to the pleasure and relaxation of another part—it carries insult to the point of injury. Professional fraternization on an informal level is the leaven which in many cases, if not most, raises professional relationships to that point where they become comfortable and satisfying; without it they tend to remain stiff, uneasy, and coldly impersonal. Where encounters are kept wholly in the job context, to be arranged by appointment or through channels, they are likely to remain encounters, formal and unprofitable. The lunching arrangements at King's virtually insured that, for women staff, encounters with their male counterparts were formal and unprofitable, and that such arrangements existed at all said a good deal, implicitly, about the status assigned to women, not one that could be described as equal.

No woman would have needed to be a doctrinaire feminist to be aware of this. In a perfectly practical sense, Rosalind found it difficult to know her male colleagues; but one can hardly doubt that she also resented the care with which they were guarded against her acquaintance. It was nothing, after all, with which she was familiar. From the time she left Cambridge she had had the good luck to avoid exclusionist traditions: CURA was too young an organization to have developed any, and in Paris the notion of them did not exist. But it is more important to notice that the exclusionist tradition represented to her a severe deprivation. Rosalind liked professional fraternization; it may be said that she thrived upon it. Not all of her friends were scientists, but many of them were, and it was with these that she had her closest, most complete relationships that were intellectually satisfying as well as emotionally sustaining.

The camaraderie which prevailed at the laboratory on the quai Henri IV was evident even to the casual visitor whose fraternization had nothing professional about it; to Rosalind it represented something of an ideal. This camaraderie was extensive. It embraced the storms of temperament which sometimes raged; it projected itself into a considerable social life which extended outside working hours. Even apart from the fact that it supplied Rosalind with a pleasant existence, it provided her with the sort of companionship she liked best, profited most from, was happiest with—and which, because she was by no stretch of the imagination naturally unsociable, was of enormous importance to her.

It was exactly this which she lacked at King's. She may not have imagined finding substitutes there for people she missed —for Mathieu, for whom she felt a kind of daughterly affection, or for Méring, her *cher collègue,* or for Vittorio Luzzati, her *bon copain*—but most certainly she had not anticipated a situation of doors closed to her. For Rosalind, this was a first and fundamental disappointment; it also formed, objectively, part of the surrounding circumstances.

What is interesting is that few, if any, of the people who were in Randall's laboratory at King's at the same time as Rosalind were aware of how the circumstances there appeared to her, and that few have given thought to it since. But it is necessary to add that these are men, recollecting what was a man's world, comfortable and friendly.* That Rosalind was exceptionally lonely in it never crossed their minds.

It is also interesting to notice that, if Rosalind made no friends at King's, she did not acquire enemies either. She came into conflict with no one except Maurice Wilkins. So it must be assumed that she was not so affected by loneliness, or unhappiness, or a sense of exclusion, that she became literally, or generally, impossible to get along with. Her working relationships, with a single exception, were amiable; but in the one instance which was not, it was unamiable indeed.

Why? This is not really knowable. It is a mystery; but it is not an insignificant one. It may come as a surprise to those who think reverently of science as something done by people of remarkable and inhuman detachment to discover that the course of scientific discovery in our time was much affected by the human inability to remain detached. But this is precisely what happened, and because it did, nothing is exactly the same as it would have been otherwise. It would be nice to know, then, just what factors in the human equation so affected history; but no one kept a journal, more than twenty years have passed, half the evidence is unavailable in any case, and what remains is the kind provided by recollection. It appears, from a distance and at a guess, that Rosalind and Wilkins got off on the wrong foot, and that neither of them was wholly responsible for this.

There was some misunderstanding from the start, probably a subtle one, probably also a minor one of the kind that nobody likes to make a fuss about. Every organization has its flaws, and laboratories have organizational problems which other institutions do not suffer from, for few of them are susceptible to neat, businesslike arrangements, and those which

---

* At the time there was one other woman scientist on the laboratory staff besides Rosalind.

try to impose the business concepts of chains of command and responsibility generally fail as laboratories. But there are moments, even in research, when a few definitions come in handy. It seems never to have been clearly defined what Rosalind was to do at King's—which would not have mattered, of course, if such general friendliness had prevailed that definitions were unnecessary. But Rosalind had her own idea of what she was there for, Wilkins may well have had a somewhat different one, and the uneasiness naturally produced by such differing notions was not soothed, or clearly resolved, by Randall, who was very probably unaware of the uneasiness until it had developed into a good deal more than that.

In her application for the Turner-Newall Fellowship which Randall made available to her, Rosalind stated her wish "to be in contact with a strong biological group as well as an X-ray diffraction group highly specialized in work on subcrystalline materials." The phrasing is confusing, in that only the strong biological group existed at that time at King's; the X-ray diffraction group had yet to be created. According to Randall, it was his intention that Rosalind create it; and certainly this was what she set about doing. That there was no such group in existence when she arrived is indicated by the fact that the first thing she undertook was the obtaining—in some cases, buying, in other cases, making—of suitable equipment.

There were, of course, a number of research projects going on at King's and, no doubt, many of them could have benefited from the kind of investigation offered by X-ray diffraction; but it is possible that DNA was the most likely candidate for Rosalind's attentions. Before she arrived, some attempts had been made at X-ray diffraction photographs of DNA (recent attempts, that is; W. T. Astbury had made photographs much earlier) without very promising results, which is not surprising in view of both the lack of experience in the experimenters, and the simplicity of the equipment. Surely it was intended that Rosalind should try to improve upon these results, which was, of course, exactly what she did. But the question of in whose province DNA research lay seems not to have been resolved. There may have seemed no need to resolve it. Several people at King's had worked or were working on various facets of the DNA problem, including Randall himself and, of course, Wilkins. Certainly it would have been absurd to exclude Rosalind from it, the more so as the techniques which she could bring to bear upon it were the ones that held out the greatest immediate hope of success. But it is perhaps unfortunate that Wilkins was not present at a conference held on the subject of Rosalind's participation in the DNA problem which took place very shortly after she arrived.

No one seems to recollect much about this conference apart from Raymond Gosling, who was a graduate student at the time, and who had been doing research on DNA, who had, in fact, made some of the X-ray diffraction photographs done before Rosalind came. At this meeting, he was detached from his previous work and turned over to Rosalind—"the Ph.D. slave-boy handed over in chains, so to speak" [2]—and, indeed, from this came one of the pleasantest relationships Rosalind had at King's, for she and Gosling got on very well. But Gosling does not remember that any particular lines of demarcation concerning the DNA work were ever laid down, and in this Randall agrees with him, suggesting that there may have been from the outset some "misunderstanding" about who was to do the "primary DNA research." [3]

That Wilkins was away when this conference took place probably made it injudicious, for in his absence the question of the fundamental relation of his work and Rosalind's was not raised, and he may have been somewhat surprised when he returned to find that DNA had passed at least in part into her hands. But if all else had gone well, surely this would not have mattered much. All else did not go well.

What Rosalind and Wilkins had in common were the simple things—sometimes sufficient to launch a friendship, and sometimes all that is needed to sharpen a dislike. They were close to the same age (Rosalind was four years the younger) and they had both been educated at Cambridge, where their terms of residence had overlapped, though they seem never, during that time, to have met. Their backgrounds were not identical, but they were enough alike so that it may be presumed that they spoke much the same language. Their professional experience was dissimilar, but in some cases this would have signified only a pleasant lack of direct competition, and an opportunity to exchange information drawn from different, but possibly related, sources. Wilkins had moved into biophysics from physics, after some years spent working first on the Manhattan Project, then later at the University of California on the separation of uranium isotopes by mass spectography. Rosalind was entering the same field as a physical chemist who had added a specialized kind of X-ray diffraction work to her skills. Each had something to learn from the other, and there seems no obvious barrier which should have prevented the exchange of knowledge out of which, in so many cases, scientific advances are born. But they seem in the course of more than two years never to have achieved so much as a simple conversation. It was indeed a matter of personalities clashing.

Yet they had characteristics in common. Wilkins was a sensitive man; he still is this. He is an attractive man, rather shy in a way that suggests vulnerabilities. Tact is often required in dealing with people of this nature, for they are likely to respond to affront with silence and withdrawal, with noncommunication. Their form of aggressiveness is usually quiet. Rosalind was also sensitive and shy, but she went to some lengths to conceal both, as if they seemed weaknesses in which she refused to indulge. To most people she appeared confident, entirely capable of protecting herself, and only those who knew her well came to understand that the manner did not express the true content.

Some ordinary conversation, some neutral exchange of commonplaces, some casual contact now and then in undemanding situations, might have contributed a little of what was required to explain Rosalind to Wilkins, and Wilkins to Rosalind. This seems always to have been absent, and the lack was a serious one. Rosalind and Wilkins differed very strongly in "style," and nowhere was this difference more marked than in their individual approaches to professional subjects. There is, for instance, a kind of hot and heavy form of disputation which many scientists enjoy; Rosalind enjoyed it, and found it useful. Wilkins disliked it very much, and has said so. Rosalind approached science with passion, with fire, powerfully. Gosling describes her manner of proceeding, and quite fairly, as "a very sharp debating style of discussion. . . . If you believed what you were saying, you had to argue strongly with Rosalind if she thought you were wrong, whereas Maurice would simply shut up. He wouldn't really go out on a limb and justify himself. . . . Rosalind always wanted to justify herself, or, if she was discussing with me, she always expected me to justify myself very strongly indeed. And of course I found this a tremendous help. I learned a lot from her that way." [4]

As a teaching method, strong debate has much to recommend it; we know from Socrates how effective it can be. But if students are capable of accepting the demands of a sharp debating style with equanimity, and regarding it as an opportunity for learning, those who are not students are often less comfortable with this approach. Moreover, to the English, and particularly to English intellectuals, it has a foreign quality, for it is a commoner way of going about discussion on the Continent than in Britain—one can see it being carried on, concerning any topic under the sun, in virtually any European café. That it was natural to Rosalind's temperament, and had been polished and encouraged during her years in Paris, did not make it seem a less formidable habit. No doubt Wilkins's inclination was simply to shut up. For all that he and Rosalind had in common, the gap between them was very nearly unbridgeable.

It was most unfortunate that the first recorded contact between the two was one that may be described as a clash. Exactly what took place is hard to ascertain, but in the circumstances, it is reasonable to prefer Wilkins's version. When Rosalind arrived at King's, the group working under Wilkins's direction had been for some time attempting to hydrate fibers of DNA—a sensible procedure, as many materials exhibit very different characteristics, and possibly informative ones, in a hydrated form. They had had little success, for DNA is among those substances which do not take up water easily—the procedure is by no means as simple as merely soaking one's specimens for a time in the hope that the water will penetrate. Rosalind promptly solved the annoying hydration problem with the simple suggestion that appropriate gases be bubbled through the water bath, and this, in fact, worked very well. There seems in this no room for conflict. A problem existed which had been solved, and that might well have been the end of it; but all the same, at precisely this point, two personalities came into conflict.

Wilkins says that Rosalind "took a very superior attitude from the beginning," and points out that there was nothing original, nothing inventive, in Rosalind's suggestion.[5] And this, indeed, is absolutely true. To Rosalind, that may very well have been the point. The technique for hydration which she recommended was *not* new; physical chemists say that it was a well-known and standard method described in textbooks. If she had been required to suggest something obscure, or something which demanded unusual inventiveness, she might have been more sympathetic and less "superior"; but what was—at least, relatively—common knowledge she expected, perhaps naïvely, to be commonly known. Gosling puts it very well: "She was firmly convinced that you pursued a problem because you were interested in it. You then picked up any existing techniques which enabled you to get an answer. . . . If it meant learning how to skin-dive, then she'd have learned how to skin-dive, it was as simple as that." [6] In Rosalind's mind, it was as simple as that, and beyond a doubt she was habitually stern in her judgment of those who did not appreciate this simplicity.

In retrospect, it appears that there could not have been a worse beginning. The result of this initial encounter over a problem, and a fairly minor one at that, was—from Wilkins's point of view—a strong and probably ineradicable impression that Rosalind was difficult, by nature "superior" and overbearing, for all that what she produced as a suggestion was unoriginal. To Rosalind it appeared very differently. Time should not be wasted on minor problems to which easy solutions existed, and if one happened to know the solution, to indicate it was obligatory. But because her suggestion was uninventive, to her way of thinking generally known, certainly easily obtainable knowledge, she was amazed that nobody had sought it ought. Because it required no great gift beyond a willingness to consult the literature, or a physical chemist, she was skeptical of the seriousness with which the problem was being attacked.

Her respect was never, all things considered, very difficult to achieve, but it was also not difficult to forfeit. Gosling has also commented upon this quality in Rosalind with experienced perception, "She didn't suffer fools gladly at all. You either had to be on the ball, or you were lost in any discussion about anything, and that was constant." [7] Gosling did not object to this; he describes it as "an interesting experience for a research student to have," from which he learned a great deal.[8] But he was, of course, a student. Students are noted for their ability to remain untroubled when those who teach impose rigorous demands upon them, for their willingness to accept much hard challenge without taking any of it as personally offensive. This is a willingness which often vanishes with time. For many of us, seniority is equated with authority, and neither likes challenge. But it never vanished in Rosalind. Authority remained a topic of debate with her; she never minded being challenged, and she assumed in a rather headstrong way that what she did not mind was universally unobjectionable. More than this, she saw challenge and the response to it as the method best suited to intellectual exchange, and she was passionate about it, which is reasonable, for it is to begin with the method of a passionate nature.

Certainly it was a quality that one either liked and appreciated, or one didn't. And there were those who didn't. Wilkins remembers Rosalind as "very fierce, you know. She denounced, and this made it quite impossible as far as I was concerned to have a civil conversation. I simply had to walk away." [9] Wilkins is consistent in his dislike of this passionate method; he does not enjoy talking to Vittorio Luzzati either, whose technique of discussion is not very different. Rosalind and Vittorio were warm friends; if they threw down challenges to one another like so many gauntlets, their arguments were often productive, and no offense was intended—or taken —even at the most denunciatory-sounding moments. It might, in fact, have been helpful if Rosalind had realized how general Wilkins's dislike of fierce argumentation was, for she suspected at times that it was not so much argument that he abhorred as argument with a woman.

The initial incident having to do with the hydration of DNA fibers was a small one, for nothing much depended upon the way it was resolved apart from the future relationship of two people. Of course, in this sense, it was not small at all. Plainly Wilkins was sensitive to what he saw as Rosalind's "superior attitude," believing that it indicated contempt; his refusal to join in the sort of exchange she found natural and useful she interpreted, though perhaps not so quickly, as a sign of his contempt for her.

No doubt there are many people who are capable, even after so bad a beginning, of patching things up, making a fresh start, forgiving and forgetting. These were two who were not. There was a good deal of inflexibility on both sides, and no one— which is most regrettable—chose to intervene, to take the responsibility for making a peace, except perhaps Gosling, a cheerful young man who liked both Rosalind and Wilkins, but who was rather severely limited in what he could negotiate simply because he was young, and a student, and unequipped with any powers whatever beyond those of persuasion. If Gosling managed to arrange a few temporary truces, a few relatively peaceable exchanges of views, it was the most he could do. The fundamental relationship never improved; indeed, it steadily grew worse. And the consequences of this can only be described, more in sorrow than in anger, as regrettable in the extreme.

Of course, it was all foolish. But a good deal of human behavior is, and scientists are not more free from folly than other people. That Rosalind suspected that she was ill-received at King's, and especially so by Wilkins, because she was a woman is no more odd or illogical than that sensible men of good intellect, and scientists among them, imposed and maintained a tradition of segregation by sex, and in so doing went to some lengths to insure a lack of easy, natural, comfortable, and sociable contact with their female fellow workers. If Rosalind persistently saw Wilkins as a man who did not very much like to have women around, this was no more unaccountable in her than it was in him never in his career to have taken on a woman student, though there must have been one or two that deserved the honor.* If Rosalind's style offended Wilkins, it was childish of him to deal with this by huffiness; certainly it was tactless of her to persist in eliciting the huffiness. That scientists are often enthusiasts tends to make them a little childish, and because the closest emotional interplay most of them experience is between themselves and their work, they are sometimes a little obtuse about other sorts of interplay. In this case, foolishness did not quite end with that, and this is the pity.

To this day it has not ended, and that is an injustice.

---

* Up to 1971, all the students supervised by Wilkins at King's were men.

SIX

The Making of
a Discovery

What was going on between 1951 and 1953 was a race, with the discovery of the structure of DNA as both the goal and the prize. This sort of contest is far from unfamiliar in science, which is a competitive business; what is odd about this instance is that some of the contestants did not know that they were racing. Rosalind was one of them.

The nature of this race is less well-documented than might be expected, much less than is usual. Modern science keeps close track of its own progress, recording itself in those innumerable journals in which scientists publish their findings: experiments, discoveries, conclusions, perceptions, and, occasionally, even their failures, lack of conclusions, bewilderments. These journals do more than simply record history. Any field of science is an intensely, and often a very rapidly, cumulative subject, the substance of which is constantly changing and expanding; and these changes in substance are absolute. In other words, any major new published discovery has an instant effect that is profound, that may require work in progress to be abandoned because it has been shown to be based upon a wrong guess or a disproved assumption, that opens out immediately prospects of new investigations, that in no case may be ignored. There is no element of choice in this either. A doctor who keeps up with the medical literature is not obliged to incorporate into his practice everything he has read about. Provided that most of his patients survive, neither the success nor the value of his work will be judged by how up-to-the-minute or innovative his methods are. But those who do research lack the option to be old-fashioned; the success or value of what they do will, indeed, be closely and invariably linked to whatever the newest relevant information on the subject may be. The published literature of science is, in fact, the subject itself, history in the retrospective sense, but also the expression of the current sum of knowledge at any given moment.

Because of this, most discoveries emerge, not out of the blue, but out of a clear line of progression, traceable in the literature. It is usually possible to perceive with some accuracy who contributed what, even to a solution that took a long time in the finding. Though opinions may differ as to the relative importance of various contributions, there is rarely any doubt of their existence. But when we come to DNA, things are not quite like this. The literature does not tell everything, and we know that it does not, because J. D. Watson wrote *The Double Helix* in order to tell us so.

We learn from *The Double Helix*, for instance, that a race was going on. Indeed, two races are mentioned, though one of them—the one against the devilishly clever brain of Linus Pauling—was apparently rather illusory; [1] the other, however, was real enough. King's College (London) was pitted against Cambridge, or much more accurately, Rosalind and Wilkins were vying with Watson and Crick.

It was this of which Rosalind was not entirely aware. She was not a fool, of course. She did not live in an ivory tower either; she was a practicing scientist, and an ambitious one, and she was perfectly conscious of living in a competitive world. It is true that before the structure of DNA was worked out, there was no way for anyone to know that it would prove to be of almost unique importance; but certainly it was understood that the problem was a significant one, and Rosalind did not for a moment imagine that this understanding was confined to King's College. Had she, however, chosen to make out a list of rivals, it is probable that Watson and Crick would not have been on it. And this is a very unusual situation.

Scientists are communicative people; they are obliged to be. The obligation also exists in the sense of a duty requiring them to publish their findings. In science, even more than elsewhere, to suppress a truth is to consent to a lie. But science is also competitive; and because of this it is burdened with a problem that is not easily solved. If half the motive behind the duty to publish one's findings is a duty toward oneself, the other half is an acknowledgment of the necessity of pooling information and knowledge for the sake of science itself. Indeed, if this were not done, and new truths were kept secret, the progress of research would slow to a crawl, if only because time and energy and the resources of intellect would be devoted to repeating what has already been done, rediscovering what has been discovered, duplicating what already exists.

Scientists are immensely sensitive to their urgent need for free communication; for this reason, there have been many of them who have objected to the restrictions placed on exchange of information by governments which like to keep some areas of research veiled. There are few who are not uncomfortable, to say the least, when required to compromise their right to publish which is also their obligation. On the other hand, the duty to keep telling, to keep providing new truths for the benefit of others, is an unnatural one in terms of a highly competitive society. It does not exist in commerce because commerce is pure competition. A new engine invented by Ford will not be confided to General Motors; there is no moral obligation which urges that it should be; and, if General Motors has wasted a great deal of time and labor and money producing cars which cannot compete with Ford's new line and the value of its shares consequently declines, that is the nature of economic life in a capitalist and competitive society. Nobody complains.

In commerce, the motive is happily single: to make a profit, if possible. But it is not so simple for scientists. Science, after all, is not simple. As a social activity—divorced for the moment from its technical content—it is humane in the sense of aiming to provide useful knowledge for the general benefit, and not for profit. But this end is reached through means which involve as much competition as commerce does. Scientists too must eat; how well and how often they eat depends to some degree upon their making a good showing in one race or another. More than that, scientists, like most other people, have egos which need to be satisfied, which rejoice in praise and reverence, which cherish secret dreams of immortality—Newton's Law, Einstein's Theory, Darwinism, why not add another name to the list of the unforgettable?

The urge to compete and the need to communicate are in opposition, and the balance which any scientist must maintain between the two is a delicate one. Because scientists are dependent upon communication—and not only the printed word; on their own topics they tend to be obsessive talkers—they need definitions sufficiently agreed upon so that the communicating can take place in an atmosphere of some trust. This is particularly necessary because the most interesting and valuable communications are those with people who are at least potentially competitors, who are concerned with the same problems, who share the same body of knowledge. What must be known is the extent to which the person talked to is really a rival; what must be understood is the extent to which anything divulged, or developed, or discovered in the course of the communication will be appropriately credited.[2]

Such agreement is not thrashed out anew at the beginning of every conversation; it does not need to be. It is assumed, because there is a body of practice, etiquette, manners, which is generally subscribed to, and which covers most cases. It is based upon the respect for priority of publication which governs the rewards science gives to her own—a respect which is not exactly copyright or patent, but which all the same confers a kind of right. Those named as discoverers are those who published first; on priority of publication rests the right to acclaim. Because Mendel published his discoveries very obscurely, they were rediscovered rather than communicated; but though Correns and de Vries reached the same conclusions that Mendel did, and quite independently, the credit for originality is not theirs, but his. This looks like a nicety, considering the circumstances, considering that Mendel was a long time dead and that his researches had been uninfluential, considering that those who rediscovered his work did so in ignorance of what had gone before. But it is more than a nicety; it is, indeed, the moral equivalent of copyright or patent; and if it is only moral, that does not lessen the claim, for it is not in any case ever carelessly to be assumed that either the desire to be credited with one's accomplishments or the right to that credit is always tied to profit.

Because there is no profit, no legal claim, and because the output of scientists is not only freely borrowable, but is intended to be borrowed, these scientists must either trust each other considerably, or else maintain so discreet a silence that the progress of research is impeded. On the whole they trust each other, and are right to do so. Candor is the chastity of scientists, and generally, it is diligently preserved.

One occasion for candor is the frank announcement that one has entered a race. This announcement was never made in Cambridge, nor received in London. There are circumstances which explain this; whether, as things developed, they entirely justify it is another question. There existed what might be called an administrative understanding that placed the DNA problem in Randall's laboratory at King's. Such arrangements are not ideal. Ideally, all problems should be available to all comers on an open, competitive market; and so most of them are. People in Dallas, Liverpool, Berlin, and Peking may all be grappling with the same puzzle at the same moment, and probably this is exactly what is happening. Rivals may well be operating out of laboratories in Berkeley and Cambridge, Massachusetts; and so they should be. Rivalry is stimulating and useful, and this is the way in which science works.

At least, this is the way science works most of the time. But the free-market approach is an expensive one which produces a considerable amount of duplicated effort. Therefore, there are sometimes situations in which efficiency suggests the avoidance of competition. If there are three problems urgently in need of investigation, then it is more sensible to turn over one to each of three laboratories, rather than having all the laboratories turning their attention to the most attractive problem, while the other problems languish. This sounds very good on paper; where the resources which finance science are limited, it is an appealing approach; and in more than one place and at more than one time it has been used, sometimes by formal agreement and sometimes by informal understandings.

This is an approach which scientists themselves dislike, and they tend to resist it, except under extraordinary conditions, such as during a war, for instance, when priorities intervene which need not have much to do with science itself. And they are probably wise to resist it, for research is too creative a business to profit from being narrowly channeled. But sometimes it happens. One reason why those at King's College did not realize that people at Cambridge were working on the structure of DNA was simply because an understanding existed that they weren't.

This understanding was acknowledged. For that we have Watson's word. He points out in *The Double Helix* that molecular work on DNA was, in 1951, essentially the property of King's College, and in particular, of Maurice Wilkins, and deplores the coziness of England which made it awkward for one scientist—in this case Francis Crick—to move in upon territory claimed by another: Wilkins. Though this might have been acceptable with respect to a foreigner, "the English sense of fair play would not allow Francis to move in on Maurice's problem. . . . In England . . . it simply would not look right." [3]

The statement Watson makes may represent a partial misunderstanding of the state of affairs, for as he puts the case, it is not very convincing. He adds that it would not be reasonable to expect someone at Berkeley to ignore an important problem simply because someone at Cal Tech had started on it first; [4] but then, one does not really expect this in England either. English scientists are no less scientists than those in other countries; they are no less ambitious; and whatever they may have learned on the playing fields of Eton or Manchester Grammar School does not leave them so obsessed with notions of fair play that they hover outside the laboratory door, incessantly murmuring, "After you, old chap."

This would be absurd; as Watson has put it, it is absurd. No reasonable person would consent to such nonsense; and this unreasonableness may be what Watson is suggesting. But England, in 1951, had not vast sums available for the financing of research, unlike the United States, which may have been the image Watson had in mind. Not so much a notion of politeness as a need to make the money go around had produced a certain amount of that resentable practice of dividing up the problems, on an informal level, evidently, and not by any bureaucratic dictation. Those who head laboratories, and want funds for which they must often apply to government agencies, are well-advised to base their applications upon a list of projects not also going on elsewhere; otherwise they may be asked to explain exactly why it is that their organization should be provided with support in order to do the same thing as someone else.

In these circumstances, it is not unusual, and it is sensible enough, for those who head laboratories to communicate with each other and to reach some agreements intended to avoid duplication. Such an agreement—unenforceable, informal, very possibly in the noblest scientific sense undesirable—existed between the administration of the laboratory at King's and its opposite numbers at Cambridge. The reason for making it was neither stupid nor absurdly punctilious. But it was an embarrassment. The Cavendish Laboratory at Cambridge had its problems to work on; the biophysics laboratory at King's College had others; and what King's had included DNA. To announce openly that a race for the DNA prize had now begun between the two institutions would have violated an agreement—admittedly, a gentleman's one—and not to announce it was contrary to that complex understanding on the basis of which scientists can manage to be both rivals and trusting friends.

These were the horns of a very real dilemma. But, then, the race was not really between the two institutions. Sir Lawrence Bragg, who directed the Cavendish, was by no means inclined to encourage the investigation of DNA under his roof. On the contrary, according to Watson, what Bragg said was that he and Crick must give up DNA,[5] and apparently—again, according to Watson—Bragg held firmly to this decision until the work that was going on anyway, in spite of his fiat, had reached a point at which no scientist could advocate, or even tolerate, its suppression.[6] The race was, then, a personal one, between Crick and Watson on the one hand, and King's College on the other. This is a fact of great importance.

No one at King's College seems to have realized what was going on. That there were people in Cambridge—and in a number of other places—who were interested in DNA, yes; that they were developing their own ideas and techniques, yes. But that there was a hell-for-leather gallop for the finish line, toward the independent publication of a discovery, no. Rosalind did not know, but she was not in contact with Crick at all during this time, and not much in contact with Watson, whom she disliked. Wilkins appears not to have known, although he saw a good deal of Watson and something of Crick—who was in any case an old acquaintance—and his relations with both were amicable. Randall certainly did not know. Ought they, individually or as a group, to have suspected? Hindsight says yes; but at the time such suspicions would have seemed faintly paranoid, when, after all, it was understood . . .

No one suspected. Whether this ignorance of the true state of affairs made the least difference is an arguable point. To know that other people are scrambling after the same prize can be something of a spur, but only within limits. Rosalind, for example, was so much in the habit of pushing herself to the full extent of her energy and application that competition could not have elicited a great deal more than was already forthcoming. Nor was the situation one in which recognition of rivalry might usefully have produced something more in the way of money or support, because neither was required. What awareness might have produced was, perhaps, discreet silence, or possibly—but this is pure speculation, of course—some lessening of the friction between Rosalind and Wilkins by providing a common cause in which they might have been able to join sufficiently to overlook what each found to be failings in the other. Nobody can say that the conversations they might have had, had they been able to converse at all, would have been significant and productive. All that can be said is that neither felt any external pressure that required them to try.

Largely because of this situation, although Rosalind came very close between January 1951 and March 1953 to solving the DNA problem, she was beaten in the end by Crick and Watson, who had in their success more help from her work than she ever knew they had received.

How can this be? Science is endlessly communicative. Scientists publish, and what they publish is public property. Anyone can seize upon an idea, a method, a result, a perception, a theory, appropriate it to his own purposes, and carry on from there. All that is required is the small, proper note of acknowledgment which makes perfectly clear who contributed what. Scientists also talk. They give lectures and seminars; they have conversations. When they write, however, what they write is not always for publication—apart from correspondence, there may be reports on work in progress not intended for immediate circulation, containing material, for instance, not ready for public scrutiny and criticism, or perhaps the raw data from which a publication will one day emerge. (And it is important that there should be no pressures leading to premature publication merely for the sake of claiming the problem.)

Scientists write; they talk; and when they talk, what they say is sometimes on the record, sometimes it is not. No professor would dare to lecture on his unfinished or unpublished work if he had to assume that some listener would transform his teaching into a paper that would secure to someone else the credit for the notion or the result. No two people working in the same field could enjoy shoptalk, a chat, a good argument, if each did not feel bound by the same limitations. Talk is essential, talk stimulates, arguments clarify, speculations which are thrown out to the winds may fall like seed to spring up with a crop of perceptions.

That very few scientists are really taciturn in the presence of others in their special field is understandable, they cannot afford to be. And they do not need to be. This inveterate chattiness is no threat, because it is carried on within limits prescribed by convention, on the understanding that if brains are being picked, then a proper and sufficient acknowledgment of this will turn up in due course. Brain-picking can be so mutual, indeed, that a good many fruitful collaborations have been unplanned, being no more than the result of one person's musing to another, who muses back, until out of this meeting of two minds a joint work is produced. More frequently what occurs is an exchange of profitable suggestions, usually confessed to in that last paragraph of published papers where the acknowledgments lie. That the confession is made is, of course, a matter of etiquette; it is unenforceable, except in the area of reputation. But let it not be thought that good manners are unimportant. When we apply to them the name of ethics, we are not elevating right behavior to a higher plane than it deserves, but only recognizing how essential to community survival right behavior is.

Circumstances which discourage right behavior are to be deplored. The dividing up of intellectual problems into private preserves not to be poached upon may well be a deplorable practice because it discourages free intellectual activity on the one hand and on the other discourages candor when intellectual activity has asserted its freedom. Two people taking a great "unofficial" interest in a problem "offi-

cially" not theirs to attack can, no doubt, find candor hard to
achieve. It appears that in the case of DNA something close
to this occurred. There is reason to suppose that candor will
never be achieved in this case. The record has, quite simply,
become extremely unclear. There is some mystery where
there ought to be history. There is insufficient documentary
evidence. Too many of the events which led up to Crick's and
Watson's publication of their monumental discovery can be
reconstructed only from reports of what people think they
remember; these reports conflict, and sometimes conflict enormously. This is not to be wondered at. Memory is never very
reliable, and at a remove of twenty years or so from what it is
recalling, it is rarely to be sworn to; this is true even when
what it is dredging up out of the past is neutral material, not
calculated to trouble anyone, and this is not a neutral instance.

The conflicting reports indicate a lack of neutrality; sides
have been taken, and however multiple and shifting these
may be, loyalty no doubt forbids dispassion. It is not—and
this should be clear—that anyone can be said to be lying.
When an atmosphere grows thick enough with justifications,
explanations, rationalizations, postures, and regrets, not to
omit occasional hostilities, untruth disappears just as surely as
truth does. What remains is a series of viewpoints, none of
which can be taken as proof of anything more than the state
of mind of whoever expresses them, always speaking sincerely enough, beyond question, but not precisely with the
cool calm voice of historical accuracy.

Much is lost, and this is a pity. It is not less regrettable
because it is not a unique instance. Fact has been devoured
by opinion before this. That is why we do not know much
about Richard III, except for rumor. Lost facts are not always
replaceable; certainly opinion does not replace them, nor does
a legend. Nobody involved in either side of the DNA work
kept a journal and, though the existence of a contemporary
written record is often rumored, none of the people to whom
this diary is attributed has ever admitted to its existence; certainly no one has produced it. The only approximation of a

running account is in a set of notebooks which Rosalind kept, meticulously entering the course of her experiments. But these are not enough—they are a scientist's daily logbooks, testifying to a neat and precise turn of mind, but containing no personal or unnecessary comments whatever.

Among all the opinions which exist—and surely this is odd —Rosalind is the one person involved in the events whose point of view cannot be reported, for the overwhelmingly simple reason that she never had one. The events which have subsequently been the subject of so much curiosity, so much speculation, which have been so much explained and justified, were ones of which she was unaware. She was, in fact, profoundly innocent, she never asked, never guessed, never was told.

SEVEN

"She Was Definitely Antihelical" [1]

"During 1950, M. H. F. Wilkins suceeded in obtaining well-oriented fine fibres from a specimen of desoxyribose nucleic acid * prepared by Professor Signer [Berne, Switzerland], and R. G. Gosling showed that a bundle of these fibres gave X-ray diagrams of exceptional quality. In January 1951 it was agreed that I should undertake, in collaboration with Gosling, a systematic X-ray investigation of these fibres." [2] So Rosalind wrote in February 1952, in a report to Randall covering the progress of her first year's work at King's College. It is a modest statement in which Wilkins and Gosling are given their due. It is modest also in that Rosalind named herself, in connection with Gosling who was then a graduate student, simply as a collaborator. Surely this makes a little puzzling the suggestion that Rosalind, in working on DNA, misunderstood the role planned for her at King's, and took over what she was not intended to do.

But what she had to do first was to set up a laboratory. King's was not well equipped for X-ray diffraction work of the

---

* Another way of saying deoxyribonucleic acid, or DNA.

sort needed for DNA.* In fact, as Rosalind's report goes on to
say, "the greater part of the first eight months was taken up
with the assembling of the necessary apparatus." [3] (It is for
this reason—that setting up is a tiresome process—that those
who do experimental research are always a little reluctant to
move from one place to another.) Certainly the delay made
Rosalind impatient; but she had begun, in any case, to feel
another kind of impatience as she came progressively to doubt
that she would ever settle in comfortably at King's.

I first heard of this in June 1951, when she and my husband,
David, and I—along with a few hundred other people—were
in Stockholm for the Second International Congress of Crystallography. It was plain from the outset that Rosalind was going
to Sweden with something like a sense of relief; certainly in a
state of high spirits, rather like a schoolgirl let off on a holiday,
and determined to enjoy every moment of it. In those days it
was possible to enjoy the congresses, which were still small
enough to be cozy and, for once, Rosalind and I shared the
same notion of a scientific meeting—that it was a social occasion
rather than a professional one, an excuse for a pleasant trip,
and an opportunity to see a lot of old friends all together. For
Rosalind, it was also a reunion with some of her friends from
Paris, which was more than sufficient reason for her to be
there, and the prospect made her irrepressible. When one of
the English contingent, known for his left-wing sympathies,
mysteriously vanished between Göteborg and Stockholm, it
was Rosalind who speculated at length upon the likelihood
that he had "done a bunk" in the manner of Bruno Pontecorvo; †

---

* The report adds precisely, "Gosling's original X-ray photographs
were taken with standard apparatus . . . not well suited to this
type of work. The diffraction pattern of DNA, like that of proteins,
is confined to rather small angles . . . and the amount of information that can be obtained from a fibre diagram is, therefore, to a
large extent determined by the X-ray-optical resolving power of the
system used. An X-ray tube having a small focus with high in-
trinsic brilliance is required."

† Less than a year before, Pontecorvo—a physicist employed at the
British atomic research center at Harwell—had ended a trip to
Europe by disappearing in Sweden, only to pop up sometime later
safely in Russia, where he has since remained.

this delightful rumor had dozens of scientists enraptured until the lost sheep turned up, rather to everyone's disappointment, having done nothing more dramatic than miss the boat train.

While the meetings were going on, Rosalind cut most of them, taking me on long tours of Stockholm, first in search of "sticky cakes"—Rosalind had a child's appetite for sweets, sugar was still rationed in England, and the Swedish pastry shops were crammed with what she regarded as "treasures"— and then in pursuit of odds and ends for her new apartment in London. We went to look at nursery schools and housing developments, both of which left Rosalind more skeptical than impressed. She thought the children in the nursery schools overdisciplined, and the blocks of modern apartments over-uniform. Vittorio Luzzati located an Italian restaurant and arranged a dinner party which I remember partly because the food was very good, but chiefly because it was a hilarious event. Rosalind was radiant with pleasure, positively beautiful, and looking, as she always did at such moments, barely more than twenty.

In the middle of all this, I took Rosalind to a rather literary reception given by the stunningly blonde editor of a Swedish magazine which had published some of my stories in translation—and which had provided thereby enough kronor so that David and I could afford the whole trip—that may have amazed Rosalind. Certainly it amazed me. The high point of the party, as I remember it, was the presentation of a new literary review put out by university students, an untitled magazine all of the pages of which were blank. "It is a joke," my charming editor explained, "intended as a subtle criticism of modern writing." I said, rather sadly, "Yes."

When Rosalind and David and I were walking back to our *pension* after this somewhat dismal event, I commented on the uselessness of having parties in order that writers could meet one another, because there is never any reason to believe that, once they have met, they will have anything to say. Well, perhaps, Rosalind agreed; but still, she said, she rather envied writers, because at least they did not have to "waste their time having stupid arguments with people." I assured her that this was a misapprehension, as writers argue all the time, but usually profitlessly, as there are no objective standards by which literature can be judged and therefore all opinions—the sound and fury to the contrary notwithstanding—amount to not much more than an assertion of taste, while scientists presumably have evidence behind their opinions. "But it isn't really that way," Rosalind said, and began to describe her difficulties with Wilkins, with whom she felt she could not communicate, and who gave the impression of disliking what she had to say.

She spoke so unhappily that it was startling to anyone who knew Rosalind, for she was not given to displaying her feelings. She did not say much, then or later—she rarely complained, and it was more in her character briefly to sum up people or situations she disliked, and then dismiss them coldly, than to discuss such matters at length. In such situations, or with such people, she was usually at some disadvantage, for she had neither the temperament nor the diplomatic skills necessary for the kind of negotiation that can make obstacles or irritations disappear into blandness. In order to negotiate at all, she needed a friend and confidant; in Paris she had had Vittorio Luzzati; later she was to have Aaron Klug; but at King's she had no one, and it may have been her feeling of isolation there that depressed her as much as anything else. That Rosalind lacked for a friend and confidant, while everyone else involved in the DNA work was amply supplied with both, comes in the end to explain a good deal.

By November, Rosalind was happier. She had produced some interesting experimental results, and it does not take very much more than this to provide a scientist with at least some temporary elation. By November, in fact, she had accomplished enough to have something worth presenting at a colloquium given at King's. And from this point onward, history exists in variant versions.

Some facts are ascertainable. What Rosalind reported in her talk survives in the form of notes which she prepared for the occasion. Because these notes still exist, one can see that they contain very nearly all the material included in the report she wrote three months later for Randall. These are documents, not opinions. They inform us that what Rosalind had done was successfully to hydrate fibers of DNA and then to show from the X-ray diffraction photographs which she made that considerable changes took place from the wet to the crystalline to the dry states. The notion of doing this was not original with her; it had occurred to Wilkins; it had been tried before; and according to Gosling, indeed, the X-ray patterns obtained by him before Rosalind came to King's showed enough interesting reflections to indicate that this was a sensible avenue of approach.[4] The difference was that Rosalind applied an improved method of hydration, and then had the advantage, previously lacking, of appropriate X-ray apparatus for the diffraction studies.

Under the influence of effective hydration, there was a change in the fiber which may be described as dramatic. Again according to Gosling, there was a moment when the hydrated fiber literally leaped off the instrument to which it was affixed. One version of history, the one according to Wilkins, is that this experiment was "a fluke," but it is difficult to see what this comment means.[5] That no one could predict that such a vast change would occur is perfectly true, and nobody did predict it; but experiments are carried out in order to provide new information, not only to confirm what has already been guessed at. Along the line of providing fresh information, this experiment was an extremely successful one. Nor was it accidental by any means that Rosalind had managed to achieve a hydration of the material. It was her intention to wet it, and she carried out the intention with considerable efficiency. When Wilkins adds that Rosalind "did not quite know what she had there," this is no easier to understand. What she had was the B form of DNA, quite different from the previously known A form, and this difference she recognized immediately. What it signified, nobody in the world was in a position to guess in the autumn of 1951.

What she did, and what she knew, is set out in her notes and in her report.[*]

_____

[*] What is quoted is from the report, because the notes are notes. But plainly they cover everything detailed in the report.

The highly crystalline fibre diagram given by DNA fibres is obtained only in a certain humidity range, about 70% to 80%. The general characteristics of the diagram suggest that the DNA chains are in a helical form.

Higher humidity gives a diagram showing the following principal features: (1) a sharp spot at $\sim$ 22A on the equator; (2) diffuse meridional arc at $\sim$ 3.4A; (3) two diffuse spots at about 40° to the meridian.

This diagram appears to correspond to scattering by individual helical units; i.e., it shows the form factor of the helical units (except for the sharp equatorial spot which is related to an inter-helical distance). That is, at high humidity a water sheath disrupts the spatial relationships between neighbouring helices, and only the parallelism of their axes is preserved.

During the change "crystalline → wet" a considerable increase in length of the fibres occurs. The helix in the wet state is therefore presumably not identical with that of the crystalline state.

. . . With decreasing humidity the crystalline diagram gradually fades away without decreasing appreciably in sharpness. This means that the basic lattice is preserved while disorder about the lattice-points increases, more and more scattered radiation passing from the sharp spots into the diffuse background.

. . . The crystalline state is associated with $\sim$ 20% weight water (in dry DNA.) But both the water content and the structural state of the DNA fibres are highly dependent on history as well as on relative humidity: i.e., there is a strong hysteresis in water uptake.

. . . The effect of strong drying is to make the crystalline state obtained on subsequent wetting both more *stable* and more *perfect*. After strong drying and re-wetting the crystalline form can only be destroyed at very high humidities. The spots on the "crystalline" diagram are much sharper when the substance was previously strongly dried.

. . . The fibre diagram of the crystalline form can be indexed tentatively on the basis of a face-centred monoclinic unit cell . . . nearly hexagonal in projection. . . . It therefore suggests that the structure is built up of near-cylindrical units.

. . . The 27A layer-line spacing is very strong, which suggests that it corresponds to one turn of a helix.

. . . Astbury's [6] density measurement . . . together with our water-content measurements, indicate 24 nucleotides per primitive unit cell, and 4 molecules of water associated with each nucleotide.

. . . The results suggest a helical structure (which must be very closely packed) containing probably 2, 3 or 4 co-axial nucleic acid chains per helical unit, *and having the phosphate groups near the outside.* It is the phosphate groups which would be capable of forming strong interhelical bonds in the presence of considerable quantities of water (leading first to the "wet" structure of independent helices with parallel axes, and ultimately to the solution of DNA in water) and would remain strong in the absence of water, thus explaining the cementing effect of strong drying. The *dry* structure is distorted and strained due to holes left by the removal or water, but contains intact the skeleton of the crystalline structure.[7]

It is a long pull to read this. But it is worth it. Here is the documentary evidence against which opinion and recollection must be measured. In *The Double Helix,* James D. Watson has provided his recollection and opinion in detail, and they offer a curious contrast. He attended Rosalind's talk, and found it, to begin with, lacking in optimism, apparently because she indicated that hard facts would come only when further data had been collected which could allow the crystallographic analyses to be carried to a more refined stage.

Whether this proves lack of optimism is debatable; it sounds rather more like a projected program of hard work, something which is usually undertaken in a hopeful spirit, combined with a commendable prudence with respect to unwarranted predicting of results.

He was also troubled that no one brought up the desirability of using molecular models to help solve the structure, and suggests that the general lack of a romantically optimistic response from the listeners as well as the silence on the topic of models, might have been due to fear of a sharp retort from the woman he inveterately refers to as "Rosy." Romantic optimism, it is true, was not what Rosalind dealt in; scientists are advised to avoid it, as a rule. As for molecular models, it is not entirely clear what mention of them at this point in the progress of the DNA research would have contributed. The dreaded sharp retort from Rosalind on that subject could only have been a statement that she did not see at the moment what basis models could be successfully built upon, and this would have been hard effectively to dispute. Information—hard facts —that cast light upon the structure of DNA was far from plentiful; Rosalind possessed as much of it as anyone; and that it was as yet insufficient to narrow down the many possibilities for conceivable model structures she knew perfectly well. Watson may have been unaware of the fact that in her graphite work Rosalind had made use of model-building techniques, but being unaware does not seem quite sufficient excuse for an assumption that she cherished *categorical* objections to the method.[8]

Watson, in fact, did not seem to get a great deal out of Rosalind's talk beyond a depressed feeling that in it he was being told by a woman to refrain from venturing an opinion upon a subject for which he was not trained. But he seems to have gained an impression that he felt worth reporting: "Maurice's mood was . . . jovial. . . . He detailed how, in spite of . . . elaborate crystallographic analysis, little real progress had been made by Rosy since . . . she arrived at King's. Though her X-ray photographs were somewhat sharper than his, she was unable to say anything more positive than he had already. True, she had done some . . . detailed measurements of the water content of her DNA samples, but even here Maurice had doubts about whether she was really measuring what she claimed." [9]

The view is attributed by Watson; it would be entirely unfair to take it literally as Wilkins's own. If Wilkins was skeptical about the accuracy of Rosalind's experiments, or her measurements, the skepticism proved to be unjustified. What Wilkins himself has said is, "I looked at the photograph— that B-form picture—and there it was, you can see the helix right there on the picture, but she refused point-blank to see it. She was definitely antihelical. This is where the real tragedy came in, because if she hadn't had this antihelical attitude, there's no doubt the solution would have been there ahead of Watson and Crick. She was very bloody-minded about it. What could I do? I couldn't even talk to her." [10]

Rosalind's notes for this November lecture say, "*Conclusion: Big helix in several chains, phosphates on outside, phosphate-phosphate inter-helical bonds disrupted by water. Phosphate links available to proteins.*" She underlined it.*

This does not in itself provide the structure of DNA. But it contains some essential clues, without which the structure of DNA could not have been determined at the time at which it was. It is true information and new, more "positive" in that sense than what Wilkins had produced before Rosalind came or, indeed, than what anyone had produced before her. It is important not to lose sight of this fact.

It is, of course, incomprehensible that Wilkins could have heard this talk, and yet continued to believe that Rosalind was so "bloody-minded" that she couldn't see the helix in the B-form picture. Nothing could be stronger testimony to the blank wall of noncommunication between them. Some tragedy is always implied when human beings can in no way communicate; what happened at King's is only a special example of it.

There was, of course, communication, in the sense that no one was struck dumb, or deaf, or blind. If Wilkins did not learn from Rosalind herself that she had concluded that the B form

---

* The notes are in Rosalind's handwriting, headed, "Colloquium, Nov. 1951." The report is typewritten, dated 7 February 1952, and is corrected in her hand.

of DNA was helical in structure, he must have heard it from Gosling. Wilkins had the friendly habit of meeting with other staff members, and research students, at an Irish pub in the Strand called Finch's, not far from King's College, and though Rosalind was not a member of the group, Gosling was. If in no other way, the news was passed on. Whether Wilkins believed that Rosalind was really measuring what she claimed, or failed—as Watson says—to believe at all is rather immaterial; what matters is that Rosalind rightly interpreted her B-form picture, and deduced quite accurately not only the presence of a multichain helix, but that the phosphate groups were on the outside. To this nobody seems to have paid much attention either.

Watson failed to hear it. He attended the lecture but neglected—as he himself confesses—to take notes. He also confesses that he misremembered what he heard, though whether this failure of memory was caused by his preoccupation with Rosalind's hair-do or his resentment at being told by a woman what her experimental findings consisted of is not indicated. This seems to have been his first encounter with Rosalind; if his recollection is to be credited, he was curiously hostile and contemptuous toward her before they had exchanged a word, and throughout *The Double Helix*, this attitude prevails, only slightly repented-of in a paragraph of epilogue.*

---

* It is sometimes quite fanciful. On pages 166–167 of *The Double Helix*, Watson writes, "Suddenly Rosy came from behind the lab bench . . . and began moving toward me. Fearing that in her hot anger she might strike me, I . . . retreated. I told Maurice . . . his appearance might have prevented Rosy from assaulting me. . . . He assured me . . . some months earlier she had made a similar lunge toward him."

Wilkins says, "Jim somewhat exaggerated. . . . I can imagine that she might have slapped someone's face, but that's not physical attack. Jim wrote a novel."

Indeed, it does seem rather exaggerated that any man would fear assault by a small, slim woman who did not even know karate. People who knew Rosalind incline to find Watson's account of this dangerous moment merely absurd, though not from his point of view necessarily unpurposeful.

Who was Watson, then, that like the deaf adder he would not hearken? He was—and is—a geneticist; in 1951 he was young, not more than twenty-four or so, recently arrived in England from the United States via Copenhagen, and determined to stay in Cambridge and work with Francis Crick, on the DNA problem if possible. In both these determinations he showed great discernment. Let there be no mistake, Watson is a scientist of important accomplishments, and in 1951 he already showed significant promise. It is very much to his credit that he apprehended how essential it was to elucidate the structure of DNA at a time when not everyone agreed with him. That he did this was not just a matter of luck, for the kind of instinct which guides a scientist to the right problem at the right time before the rightness of either can be demonstrated is a gift of a high order, never to be despised. It is also to his credit that he immediately saw the point of Francis Crick, for not everyone had that much perception.

Crick is a remarkable man; he has never been otherwise. He is immensely and endlessly creative; his vitality is boundless. If there is a reliable definition for genius, it might be said that this is what he possesses. He also possesses great charm, and something that is better than that. Watson begins *The Double Helix* with the remark that he has never seen Francis Crick in a modest mood, but that fails to illuminate, partly because modesty is not a matter of mood, but chiefly because modesty, in the sense of self-deprecation—and what else can Watson have in mind?—is both alien and irrelevant to people who are happy in themselves, in their beings, in their skins, their natures, their capacities. Crick gives every evidence of this healthy contentment. It is necessary to admire him. Rosalind admired him very much, and with good reason. She also liked him. He is enormously likable, though perhaps in a way which strong personalities are better equipped to enjoy than mild or uncertain ones, who might find themselves overwhelmed.

Plainly Watson admired Crick from the start, and certainly their alliance was a sound and productive one. Though one rather uncharitable commentator upon the events surrounding the establishment of the structure of DNA is given to describ-

ing Watson as "a leg man for Crick," this view will not stand
up to analysis. It is too serious an underestimation, both of
Watson's abilities, which are far from mean, and of what Crick
can accomplish on his own, without such devices. It is truer
to say that they had almost ideally complementary gifts, and
added to them exactly the compatibility that Rosalind and
Wilkins lacked. Not the least significant of these gifts was
Watson's knack for keeping Crick's mind fixed on the problem
at hand. Though Crick was adept at throwing off a dazzling
stream of ideas, he did not develop early the kind of dogged-
ness that led him to settle down and pursue any one of them
to a neat conclusion. As it has been unkindly said by one who
watched the process, "Jim nagged Francis, and it helped."

It would not have helped, it might even have been detri-
mental, if their personalities as well as their talents had not
been so splendidly matched. There was evidently just enough
friction between them to strike sparks, and this, where ideas
are concerned, is far more productive than blandness. They
made a formidable combination. It was also a sufficient one.
They solved the problem, won the race, achieved the prize.

It matters how this was done, not only because the discovery
of the structure of DNA marked a significant moment in human history, but because the methods chosen so successfully
by Crick and Watson are obliquely connected with what happened. The race for the structure of DNA was more than a
race between individuals; it was equally a contest between
two different methods of structure determination, between two
ideas. It is popularly believed that in science, as opposed to
other areas, ideas are held dispassionately, with serene objectivity, if only because they are ultimately subjected to the severe test of demonstrable proof. "Does it work?" scientists are
given to saying; what is implied by the question is the existence of a clear, objective answer—yes, it works; no, it doesn't.
But whether in the end it "works" or fails to is unlikely to be
evident from the outset, and in most cases a certain amount
of time and effort, often a great deal of both, must be invested
before the workability is known.

The choice of both the problem and the approach to solving it is of vast consequence to a scientist. A wrong choice in either direction can be disastrous, a waste of productive years in unproductive labor, to an end which is at best negative—in other words, to the conclusion that it doesn't work. Inevitably the particular choice made is less dispassionately followed than appears on the surface, or—in fact—than scientists are usually willing to admit. *The Double Helix* offers evidence of this. It is a book which can be read in many ways—it is by no means as ingenuous as it appears—and part of it, though not the part the average reader would be most aware of, is a passionate argument, an advocacy at times amounting almost to propaganda, in favor of a particular method of problem-solving.

The method advocated is model building, the choice which Watson and Crick adopted; Rosalind used a different one. Neither method was new, or original with the people who used it. Model building, in a primitive sense, goes back to the late nineteenth century when not only one but two alternative model structures for sodium chloride were proposed, the decision as to which was correct having to await the discovery of X-ray diffraction. Very early X-ray diffraction work involved simple structures which could often be directly deduced from the relative intensities of the X-ray reflections, and in such instances models were not necessary; later, as more complicated structures were investigated, trial processes involving hypothetical positions for the various atoms came into use, and these processes were, of course, a form of model building. Still later, the development of Fourier methods of mathematical calculation made possible such extensive analysis of diffraction data that many structures could be determined directly by X-ray methods, without resort to model building. This worked successfully with such molecules as those of penicillin and vitamin B12, the structures of which were determined by Dorothy Hodgkin at Oxford, and with other substances similar in that they will form crystals that give many diffraction effects. It worked less well with highly complicated systems, such as proteins, and still less well with fibers, which encouraged a revived interest in model building as an alternative approach.

The mingling of X-ray analysis and model building techniques ultimately proved very powerful: out of such mingling Linus Pauling discovered the α-helix. The point is that both methods have long coexisted in a mutually supportive way, and that the choice of which to use, or whether to use both, is a practical and pragmatic one chiefly related to the nature of the problem at hand.

Rosalind used both, in this practical and pragmatic way. In the case of Watson and Crick, model building was not only the approach they chose, it was the one they advocated. In *The Double Helix*, Watson's advocacy comes close to evangelical zeal. And among the many roles Watson calls upon Rosalind to play, one is that of the unconverted heathen.

> The years of careful, unemotional, crystallographic training had left their mark upon Rosalind. She had not had the advantage of a rigid Cambridge education only to be so foolish as to misuse it. It was downright obvious to her that the only way to establish the DNA structure was by pure crystallographic approaches. As model-building did not appeal to her, at no time did she mention Pauling's triumph over the α-helix. The idea of using tinker-toy-like models to solve biological structures was clearly a last resort.[11]

Now, these statements are literal nonsense. Rosalind, to begin with, had not had years of "careful, unemotional crystallographic training," or for that matter, years of crystallographic training which were haphazard and passionate. Her Cambridge education, whether rigid or lax, had been in physical chemistry; she began her working career as a physical chemist; and what she knew about crystallography she had learned on the job in Paris from Méring. She had no vested interest in any technique; for that matter, she did not even describe herself as a crystallographer.*

---

* She was perfectly clear about this. When David was about to take his viva at Oxford for his D.Phil., Rosalind remarked that most probably she could not pass the examination, never having become a "proper crystallographer."

It also must be noted that Rosalind's affection for "hard facts"—a liking not uncommon in scientists, and not often reproved—was, in this case, perfectly sensible, and just as sensible from the viewpoint of model building as from any other. Models are not built out of thin air or random inspiration; they do not come into being quite as magically as Watson appears to suggest. What model building consists of, briefly stated, is the successive hypothesizing of molecular models until one is arrived at which has its atoms arranged in such a way that the diffraction pattern—and everything else—is accounted for, never a really simple thing to do, and probably in all instances something which demands a certain flair. But it also demands a very extensive supply of information, of hard facts; if this is lacking, a model builder is less likely to produce a molecular model than a castle in the air, a wish fulfillment.*

The data required, the hard facts, are immense. In their absence no plausible model can be built, none can be verified. The arrangement of atoms arrived at as a proposed structure for any molecule must account satisfactorily for the diffraction pattern, embrace without contradiction all known chemical, physical, crystallographic—and in some cases, biological—information, and manage to defy no law of nature. Quite plainly, if nothing is known about a substance, a model cannot be built at all; if too little is known, then any model produced will be vague, uncertain, composed chiefly of hope, and insufficiently verifiable.

How much information is needed? In Rosalind's opinion, a good deal more than existed in the autumn of 1951. She had no prejudices against model building; it was a technique which

---

* The models which model builders build are literally that: they are made of wire, cardboard, plastic, or whatever is handy. It is a little confusing that those who work exclusively with mathematical methods also build models, but in this case the model is the end product of the calculations and exists to demonstrate visually the structure arrived at. It is very difficult to depict the complex three-dimensional linkings of the atoms in a molecule in anything but a three-dimensional form.

she had used to a degree in some of her carbon work in Paris, and one which she was subsequently to use in her work on viruses. She was not dedicated to it, however; and she could not have helped being aware that at Cambridge, where model building was practiced with much enthusiasm, there had been one notable failure of the method as well as a number of successes. Certainly she knew that the model builder who has not got enough to go on may well come up with an attractive, pleasing, sensible-looking proposition which is all the same unprovable, and which may even be quite wrong. Linus Pauling proposed an incorrect model for DNA. At the end of 1951, so did Watson and Crick.

In the spring of 1953, Crick and Watson produced their triumphant and correct model: the solution to the DNA problem. By that time some essential clues had come into their hands which had been provided by Rosalind's liking for hard facts. And by then they had begun to pay attention, as they had not earlier. Rosalind had a picture of the B form of DNA—newly discovered by her—in time for her talk in November 1951. By then she had measured, and measured accurately, the water content of her samples, for all that Wilkins may have suggested that she hadn't, and Watson couldn't remember the measurements after he heard them. He ruefully confesses that his failure to recollect the exact water content was most unfortunate, for what he reported back to Crick in Cambridge was misleading.[12] It was also unfortunate that he missed Rosalind's suggestion that the phosphate groups might be deduced to lie on the outside of the molecule, as in fact they do, and as she was to demonstrate conclusively in 1952.

The consequence of what he misremembered or failed to notice was the incorrect model he and Crick produced shortly after Rosalind's talk, which had the sugar-phosphate backbone located in the molecule's center. According to *The Double Helix*, they were rather pleased with their product, though the structure they proposed was disappointingly unilluminating. Wilkins was invited to Cambridge to witness the triumph; William Seeds, who worked with Wilkins, came along, and Rosa-

lind and Gosling as well. The session was opened by Crick
with an exposition of helical diffraction theory, a subject upon
which he was very expert, and went on to a description of the
model, of which Rosalind plainly did not think much. Her disdain of it Watson accounts for on the grounds that what was
proposed was a helical structure, while Rosalind did not admit
that a shred of evidence existed to indicate that DNA was
helical—a curious statement, considering that very shortly before she had presented a good deal of evidence suggesting that
the B form of DNA was exactly that. What she did object to
in the proposed structure—and aggressively, we are told—was
that the three-chain model had its phosphate groups held together by Mg^{++} ions in a way she thought unlikely, considering that by her calculations the Mg^{++} ions would be surrounded by tight shells of water molecules.

And as Watson was required to confess, her objections,
though very annoying, were not mere perversity.[13]

EIGHT

"On the One Hand a Defeat, On the Other, a Triumph"[1]

By the end of 1951, Rosalind had had enough of King's College. In March 1952, she wrote me,

> When I got back from my summer holidays I had a terrific crisis with Wilkins which nearly resulted in my going straight back to Paris. Since then we've agreed to differ, and work goes on—in fact, quite well. I went to Paris for a week in January to decide whether or not to go back. Got it all fixed up to work with Vittorio on liquids, and then thought it was probably silly after all. Somehow I feel that to take such a big step backwards wouldn't work. So I went to see Bernal,* who condescended to recognise me,

---

* The late J. D. Bernal, a pioneer in X-ray crystallography, and a scientist of very great accomplishments. He was at the time professor of physics at Birkbeck College, London.

made himself pleasant, and gave me some hopes of working in his biological group one day—I wouldn't at that stage make it clear that I wanted to go *this* year. So I think I shall probably be working around to that. (But not a soul knows so far.) Whatever one may have against the man,* he's brilliant, and I should think an inspiring person to work under. And Birkbeck College is, I suppose, more alive than other London colleges. It has only part-time evening students, and consequently they really want to learn and to work. And they seem to collect a large proportion of foreigners on the staff, which is a good sign. King's has neither foreigners nor Jews.[2]

This is clear enough; one could not want it much clearer. Rosalind had assessed her situation, decided that it was hopeless, and taken independent steps that would lead her out of it. But it is not exactly the impression one gains from reading *The Double Helix*, where Watson says,

Clearly Rosy had to go or be put in her place. The former was . . . preferable because, given her belligerent moods, it would be very difficult for Maurice to maintain a dominant position that would allow him to think unhindered about DNA. . . . Unfortunately, Maurice could not see any decent way to give Rosy the boot . . . there was no denying that she had a good brain. If she could only keep her emotions under control, there would be a good chance that she could really help him.[3]

The suggestion is, of course, that Watson is speaking for Wilkins, from whom he derived his notions of what went on at King's. It is true that Watson and Wilkins struck up something of a friendship almost as soon as they met, and that Wilkins was given—as it later came to appear, "unwisely" in Ran-

---

* This was doubtless for my benefit; Rosalind knew that Bernal and I did not get on well. Which, of course, did not make him the less brilliant or inspiring, for he was both.

dall's opinion,[4] "naïvely" in his own [5]—to confiding freely in his eager and sympathetic young visitor; but it is not really possible that what Watson writes can be taken as a genuine or accurate report of Wilkins's confidences. Whatever Watson supposed to be the nature of the hierarchic organization at King's, Wilkins himself cannot have been under the illusion that he was either Rosalind's employer or her supervisor, or that he had the power to kick her in any direction. From his point of view, to dream of "giving Rosy the boot," whether decently or otherwise, was impracticable and indeed this was the nature of the problem as far as Wilkins was concerned. There is no reason to think that he misrepresented it. As much sympathy is available to a man who is forced to put up with a colleague he can neither get along with nor dismiss as to one who refrains out of pure gallantry from ridding himself of a thorn in the flesh. Wilkins deserved some sympathy, for the tension which existed between Rosalind and himself must have been as trying for him as for her, but it would scarcely seem just to base this sympathy upon an unsupported view of him as a man too weak to resolve an unhappy situation when he possessed the power to do so.

Wilkins was not a weak man, but he did not hold, with respect to Rosalind, any sort of dominant position. In administrative terms, there was no question of it; for this we have the word of Randall, who was the director of the biophysics laboratory.[6] In professional terms, there was no question of it either. The suggestion is entirely Watson's, and a very artful one it is.

For we are here presented with a picture of a deplorable situation. The progress of science is being impeded, and by what? Why, by a woman, to begin with, one labeled as subordinate, meant—or even destined—to occupy that inferior position in which presumably all women belong, even those with good brains. This woman is belligerent, and refuses to "help" her rightful superior; her emotionality, that most damnable of female failings, runs rampant in what is surely an intellectually contemptible way. But perhaps the progress of science is also being impeded somewhat by a man as well, one too inhibited by decency to be properly ruthless with female upstarts, and so to get on with the job.

This is all very distressing, for the progress of science may be taken as an objective precious to us all. Can we possibly condemn, in these circumstances, anything that might be done —for example, by two eager young men at Cambridge—to circumvent the situation? Of course, we cannot. The DNA problem had been appropriated by King's College, but if all that was going on at King's was a particularly foolish and futile kind of civil war, then might it not be said that King's had forfeited its claim? * And if Wilkins was rather freely confiding at times, may we not suppose that he too wanted to hasten the worthy progress of science, and was seeking the best means at hand—telling all to Watson—because there was no other way of getting around Rosalind's intransigence?

It makes, in fact, a long but coherent argument. The trouble is that all this rationalization, elegant as it is, does not quite hold water. For Rosalind, after all, was getting on quite well with DNA, and the progress of science had not ground to a halt at King's.

She was getting on well, and was also ready to leave at any time. In June 1952 she wrote me, "I've seen Bernal again, and he will take me at any time if Randall agrees, but I decided it would be bad politics to talk to Randall just before going away for a month, so that's a pleasure in store for me when I get back." [7] She meant, back from Yugoslavia, where she went in May, by invitation, to spend a month visiting coal research laboratories and discussing problems with the people who worked in them.[8] From what she wrote, it can be judged that Rosalind anticipated more reluctance than eagerness on Randall's part when it came to the matter of letting her go; and in this she appears to have been right. At least it can be said that

---

* On page 173 of *The Double Helix*, Watson says that Bragg withdrew his objections to the work Crick and Watson were doing on DNA because he had no sympathy whatever with the "internal squabbling" at King's. This is an unsupported assertion. But it is part of the long and subtle rationalization that seems to form the chief matter of Watson's book.

no matter how much Wilkins was longing to see the back of her, or how pleased she might herself have been to depart, nevertheless she lingered at King's for more than six months.

If Randall was in no hurry to see her leave, he was quite sensible. She was making considerable progress. She and her student Gosling were making it by themselves. That she was unhappily circumstanced at King's was no valid excuse, in Rosalind's mind, for applying herself with less than her usual vigor to the problem in hand, and that she worked steadily and thoroughly is proved by the notebooks which she kept.

Meanwhile the relations between her and Wilkins did not improve, but that they should have done so was hardly to be expected. Their "agreement to disagree" seems in practice to have come down to as much avoidance as possible, perhaps fewer storms, but even more frigid aloofness. This was not a particularly difficult attitude for Rosalind to maintain, though no doubt she found it a depressing one. She had, in fact, no need to seethe at Wilkins, toward whom her feelings were largely, and coldly, impersonal; but of course it is very possible that cold impersonality may have been just as hard for him to endure as any number of clashes, and it may have been worse than that.* The unfortunate truth is that Rosalind did not hold him in high regard, and opinions of this order Rosalind had little skill in concealing. Most of us resent being hated much less than we resent being despised, and if Wilkins was resentful, he can hardly be blamed.

Watson found reason to be sympathetic toward Wilkins for enduring an "emotional hell," [9] and from what he writes it is clear that Rosalind has been cast for the part of its chief fiend. It might simply be noticed that what she was living with was equally far from pleasant. That she was much complained of by Wilkins she did not fail to notice, and she did not enjoy it; if she resented this, it is not hard to see why. If she felt that

---

* It is ironic that Wilkins objected to Rosalind's temperament, and that Watson reproves her inability to "keep her emotions under control" while what Rosalind objected to in Wilkins was his emotionality, by which she meant what she thought his illogical response to facts.

her work was often disregarded, she had some reason to think it. Certainly she was isolated. Apart from Gosling, and a few other juniors at the laboratory, she had no one with whom she was in regular or easy professional contact. She liked Gosling, and they had no problems communicating; but, beyond a doubt, she missed having someone of more experience than he with whom to discuss what she was doing.[10] This lack of anyone at all with whom to talk was a very great handicap; it often is in matters of research. And it might be observed that it was never a handicap from which Crick or Watson suffered, or Wilkins either, for that matter.

In spite of all this, Rosalind made progress. In view of the progress, it is not astonishing that no one was really very anxious to hurry her out the door. It ought to be noticed that between January 1951 and June 1952, all the significant forward steps taken on the DNA problem were taken by Rosalind; and if it comes to that, before she arrived at King's, not much had been accomplished, apart from the obtaining of some sharper A-form pictures, which represented an important advance upon Astbury's earlier work. Just how ardently anyone was trying is another question. There are those who think —and Crick is among them—that the importance of the DNA structure was underestimated at King's, where it was taken as nothing more than another problem and, therefore, halfheartedly pursued.

This, if true, would of course provide another reason why this essential work should have been taken out of the hands of those who dabbled at it inexpertly or lackadaisically and handed over to people better prepared to give it its due. But there is no compelling reason to think that this was a true judgment. Whether important knowledge was being languidly sought at King's is debatable; in any case, Rosalind was far from languid in the seeking.

Beyond this, another argument is very often offered which suggests that if King's College lacked appropriate ardor in the pursuit of DNA, Linus Pauling did not. Watson makes a great deal of this in *The Double Helix*, pointing out rather feverishly that there was no time to lose on the English side of the

Atlantic, because once Pauling came to see that the structure
he had suggested for DNA was incorrect, he would not stop
until he had captured the prize—indeed, by Watson's estimate, he and Crick had at most six weeks' leeway.[11] He puts
it very excitingly, but what it means is far from clear. Apart
from Watson's assertion that it was so, there is no proof whatever that Pauling was losing sleep over DNA, or was immediately prepared to make the quest after its structure a full-time
pursuit. And supposing that he was, what then?

The force of this argument is slight; its basis is imprecise.
If it lies anywhere, it is in a place where one would not normally expect to find it. No one has as yet proposed any very
persuasive reasons, intellectual or moral, which explain why it
was better, either for science or for mankind, to have the solution to the DNA problem emerge in Cambridge rather than in
California. Putting aside motives of pure patriotism, which
could have moved Crick (but almost certainly didn't) and
which could apply only in reverse to Watson, this urgency is
hard to account for, and all the harder because King's College,
for instance, also happened to be located on the specified side
of the Atlantic.

What it all comes down to is that the suggestion of Pauling's
imminent rivalry, though presented in the terms of a suspense
thriller, is a red herring. For exactly this reason, it has been
quite useful. It is no more and no less than a splendid, and
only slightly transparent, rationalization, one that fits in neatly
with that other rationalization which implies that Rosalind, as
an impediment standing squarely in the path of scientific progress, deserved to be pushed aside. Both are inclined to the
same purpose, which is the justification of what otherwise
might seem not wholly justifiable, and the excuse for what is
not easily excusable.

In the meantime, Rosalind was making considerable progress with the problem. By the autumn of 1951, Rosalind had
not only produced the B form, and identified it, but had drawn
a surprising amount of information out of her experimental
data. The determination of the density of the B form was far

from unimportant. Anyone hoping to build a model was obliged to have this information in order to know how many chains there were in the molecule; without it, Crick and Watson cheerfully produced a three-chain model.[12] The same was true of the location of the sugar-phosphate backbone, and this Rosalind also demonstrated. She thought her work was going well, but that it was still in the early stages, with a great deal more ahead to be done.

Rosalind was not prepared at this stage to build models. Wilkins might have done so if he had chosen, and he would have had available to him more essential experimental data than anyone previously had ever possessed. He did not choose to move in this direction. Instead he turned his back on the DNA problem with the announced intention of doing no more upon it until Rosalind had departed from King's, when he would take it up afresh, beginning with a repetition of Rosalind's experiments.

At Cambridge, Watson feared that having shown the three-chain proposed model to the party of visitors from King's was perhaps a mistake, in that it would dry up the source of new experimental data. In the circumstances, King's would not be so free with invitations to research colloquia for the benefit of the Cavendish, and even casual questioning of Wilkins might arouse suspicions.[13]

Rosalind, unaware of what Wilkins had in mind, and equally unaware of the direction of activity in Cambridge, turned her attention to the A form of DNA. This has been called an error in itself, but if it was, it was the kind of error chiefly visible in the brilliant illumination of hindsight. Nobody can argue a reason to leave the A form uninvestigated. If DNA existed in two forms, obviously the two were related, but what the relationship would prove to be was hardly guessable out of hand. The dramatic nature of the transition from one form to another—the B form not only leaped off the instrument on which it was mounted in order to be photographed, but had to be glued in place so that it would not spoil the pictures by moving itself out of the X-ray beam—demanded an explanation which was scarcely available without a close knowledge of

both forms. Moreover, for the systematic and methodical crystallographic approach which Rosalind had in mind, the A form
offered considerable advantages. Being the more crystalline
substance of the two, it produced X-ray photograpahs showing
many more readable reflections that those available from the
B form; and, though the evidence which existed represented a
rather knotty problem in terms of diffraction analysis, it was
not so hopelessly difficult that no useful information could be
wrung from it. To Rosalind it seemed worth trying. And she
really did very well with it, all things considered.*

Rosalind's success with the A form has not been much commented upon. In the end, the B form proved more productive
in suggesting a structure—a fact which, it ought to be said,
Rosalind also came to realize—and, consequently, her work
upon the A form has been mostly ignored, except where it had
been useful for the purpose of accusing Rosalind of being
curiously, fanatically, and immovably "antihelical" in her esti-
mation of the probable structure of DNA. What is truly curious
is that there is not a grain of justification for the accusation.
Indeed, the A form does not present immediately visible evidence of the presence of a helical structure—that beautifully
cross-shaped pattern belongs to the B form—and in the absence
of this evidence, there was certainly a reasonable possibility
that it might, for instance, contain an uncoiled, unwound helix
which perhaps sprang into shape upon wetting. This is as
far as Rosalind went.

The Double Helix lapses into grand confusion on this subject
because Watson persistently muddles Rosalind's opinion concerning the helical nature of the B form (yes, it was) with her
more tentative opinion concerning the A form (it might or

---

* One difference of great importance between the two forms was
that the water present in the B form served essentially to isolate a
single molecule of DNA, and its X-ray pattern was therefore a reflection of this singleness. The A form, being crystalline, gave much
sharper reflections, but confusingly so, for the X-ray beam was
bouncing back not only from one molecule of DNA, but from its
neighbors as well. The propinquity of these neighboring molecules
had, for all the clarity of their reflections, an obscuring effect.

might not be), and out of this has marvelous—if either careless
or disingenuous—fun.* For example, "she became increasingly
annoyed with my references to helical structures. Coolly she
pointed out that not a shred of evidence permitted Linus or
anyone else to postulate a helical structure for DNA. Most of
my words were superfluous, for she knew that Pauling was
wrong the moment I mentioned a helix." [14] This can best be
described as naughty. It combines a reasonably accurate statement—that no evidence existed up to that point which allowed
anyone to take for granted a helical structure for the A form—
with the wholly suppositious one that Rosalind had, mysteriously, something against helices, and the utterly inaccurate
one that she had leaped to a premature and prejudiced conclusion. The point is that she had not, that she was deeply
reluctant to do any such thing.†

But suggestion lingers to the effect that anyone so blindly
stupid, or so mulishly obstinate, had no real business in
messing about with science at all. This is perhaps something
more than naughty. It is resentable. Notions have been attributed to Rosalind very blithely, and with great assurance,
but there is every possibility—indeed, every likelihood—that
they originated in Watson's head rather than in hers. What he
might reasonably have argued was whether Rosalind was wise
to opt for an intensive crystallographic study of the A form,
for this is a point open to debate and something she came to
doubt herself after a time. When she did, she turned to study-
ing the B form—her notebooks for the last half of 1952 and

---

* In conversation, Wilkins also confuses these two opinions, to the
conclusion that Rosalind was obstinately "antihelical." Her notebooks do not bear out this conclusion.

† An unwillingness to jump to conclusions is, in my experience, a
thing which few scientists deplore. By its own claims, science demands objectivity rather than blind faith or guesswork. On these
grounds it rejects the loose-minded disciplines which thrive on intuition, shots at random, and the hazards of hope. Rosalind was always chary of nonobjective reasoning. At this time, at least, Watson
seems to have been much less so.

early 1953 are proof of this—where she was, in fact, much happier.

The intensive crystallographic approach was not a natural one for Rosalind, who had, in reality, very little experience in it. In France she had worked exclusively with materials— graphite and carbons—which provided only very diffuse diffraction patterns. The B form, with its pretty indication of a helix, and its general diffuseness, fell into the category of what she was used to dealing with much more than the sharper, but very complex, pattern of the A form did, and it is not easy to see why she resisted working with it for as long as she did.

Possibly she was influenced a little by her occasional talks with Vittorio Luzzati, who may have convinced her of the advantages of dealing with that form of DNA which produced the sharper patterns. Certainly this was a reasonable choice, and in statistical terms, perhaps the most reasonable. To this day it remains true that by far the greatest proportion of the structures worked out are achieved by the methodical and mathematical crystallographic approach, and to try any other approach in the 1940s and 1950s was sufficiently unusual to require a little prudent hesitation. But it does not really matter very much why Rosalind chose to explore the A form; it is more to the point that she did not grope and blunder and fail. Her application of Patterson superposition methods may have been time-consuming, but she got quite enough out of it to be able to publish, in July 1953, a paper in *Nature* * on the "Evidence for a 2-chain helix in the crystalline structure of sodium deoxyribonucleate." [15] Though at that point in time it was merely confirmatory of the B-form structure produced a few months earlier by Crick and Watson, it was not nothing.

In April 1953, Watson and Crick triumphantly published their structure of DNA. The race was over. It is a race which has been so much publicized after the event that it takes some courage to announce that it, in fact, never took place. Despite Watson's panting over the six weeks' leeway before success was

---

* With Gosling.

announced in distant California, Linus Pauling appears not to have chosen to run. Rosalind had been working away steadily; her notebooks are sufficient indication that she kept busy; but they also fail to indicate the desperate haste of someone caught up in a contest—or so occupied in racing that she was unwilling to go off for a very enjoyable month in Yugoslavia. If Wilkins was aware of being engaged in a race, he chose such an odd method of asserting his competitiveness that it becomes quite obvious that he never had any such thing in mind.

Not long after he and Watson met, Wilkins fell into the friendly habit of discussing quite freely all manner of things: his own notions about DNA, his frictions with Rosalind, and, from time to time, what research developments had taken place at King's. There is nothing unusual in this. Scientists need conversation as much as anyone else, very often more than other people, and Wilkins was simply luckier than Rosalind in finding a confidant at hand.* He had known Francis Crick for some years; Watson, who seems to have come to London more often than Crick did, enjoyed the advantage of Crick's introduction; Wilkins and Watson had much in common intellectually; and plainly Watson was sympathetic toward Wilkins's problems in areas other than the purely intellectual. He appears, at least, willingly to have regarded Rosalind through Wilkins's eyes, and in that conflict to have no doubt about which side he was on (or so one judges from *The Double Helix*, which admittedly is not a source which can always be taken literally).

In time Wilkins came to reproach himself for indiscretion, and such it may have been; but few scientists—in those days anyway—worried much about being indiscreet.[16] They had no great need to. Especially in the more lightly populated areas

---

* Or else he wasn't lucky. Rosalind needed someone to talk to, and had no one except Vittorio Luzzati, inconveniently located in Paris. She even lacked anyone with whom to discuss her nonprofessional problems at King's. She was close to her family, sufficiently so that at least two of the people who knew her at King's commented upon the relationship, but she was reluctant to discuss her trials with any of them, a reluctance she confessed to.

of science, of which structure determination was certainly one, the need for the free exchange of ideas, notions, speculations, data, flashes of brilliance, or, for that matter, complaints and grouses, was a real one. Though the rules which governed this kind of exchange were little more than an unspecified, unwritten etiquette, they were normally regarded as sufficient, and for the very good reason that they were.

When Rosalind presented, at a seminar to which people from other institutions than her own were invited, a set of very significant findings concerning the B form of DNA, she had no qualms whatever about doing this. She assumed, quite naturally, that if any who heard her talk had reason to dispute, or on the other hand to be inspired, she would be the first to hear about it. She had every reason to suppose that any disputant who had arguments to prove her wrong would not hesitate for an instant about presenting them, not merely out of a kindly interest in preventing her from making a fool of herself, but because scientists in general dislike mistakes, deeply detest the waste of time implied by the chasing of false clues or flimsy notions, and feel a strong obligation to correct error where they find it. And equally, anybody who saw in what she had to say more than she had seen for herself, who perceived a connection or implication that she had missed, who could supply more extensive information, or a new insight, or a flash of inspiration, would feel obliged to bring any or all of this to her attention, too.

Neither argument nor inspiration was forthcoming from Watson at the time. For whatever reason, the implications of Rosalind's work were lost upon him at the time. He heard about them again, with additions, early in 1953.

How this came about is something of a tangled tale. To follow *The Double Helix,* sometime in the winter of 1953— February 6 has been suggested as the date [17]—Watson paid a visit to Maurice Wilkins at King's College. But before seeing Wilkins, he had the colorful encounter with Rosalind which has been earlier described, and which filled him with dread of physical attack. There is really no use in commenting further upon this part of his story, except to say that it is used to explain why Wilkins, on this occasion, suddenly became very confiding. What Wilkins confided was, first, that he and his assistant Wilson had been duplicating some of Rosalind's X-ray work on the quiet, and then—much more dramatically—that Rosalind had had for some time evidence of the existence of a new three-dimensional form of DNA. (This, of course, is our old friend the B form, to which Watson had already been once introduced.) When Watson asked what the X-ray diffraction pattern of this new form was like, Wilkins showed him the picture.

Watson reacted strongly, with dropped jaw and racing pulse. The simplicity of the pattern, with its dominating black cross, gave instant proof of a helical structure. The X-ray picture itself provided several of the vital helical parameters, and opened the possibility that, after a little calculation, the number of chains in the molecule could be determined. The evidence for a helix was to him, and to Wilkins as well, quite overwhelming. It had, of course, been overwhelming to Rosalind some time before.

The real problem which remained, Watson writes, was the lack of a structural hypothesis which would permit the regular packing of the bases in the inside of the helix. "Of course," he says, "this presumed that Rosy had hit it right in wanting the bases in the center and the backbone outside. Though Maurice told me he was now convinced that she was correct, I remained skeptical." [18]

Some of this is a little puzzling. Watson says that Rosalind had had a new three-dimensional form of DNA since the summer; but the B form, the helical pattern, and the suggestion that the backbone was on the outside of the structure had all been produced by Rosalind not in the summer of 1952, but before the autumn of 1951. That Watson did not grasp the significance of what Rosalind said in her seminar he confesses, and quite possibly it was not until it was substantially repeated to him by Wilkins on this occasion some fifteen months later that he understood it. He may not have seen the X-ray pictures before, and plainly they were crucial to his instant understanding. But did it take Wilkins close to a year before he heard about

what Rosalind had done, or before he saw what it meant? This
seems very unlikely, and it is not made likelier by Wilkins's
own comment on this conversation with Watson.

What Wilkins says is that yes, he showed the B-form picture to Watson, and later regretted that he had done so.

> Perhaps I should have asked Rosalind's permission,
> and I didn't. Things were very difficult. Some people
> have said that I was entirely wrong to do this without
> her permission, without consulting her, at least, and
> perhaps I was. . . . If there had been anything like
> a normal situation here, I'd have asked her permission,
> naturally, though if there had been anything like a
> normal situation the whole matter of permission
> wouldn't have come up. . . . I had this photograph,
> and there was a helix right on the picture, you couldn't
> miss it. I showed it to Jim, and I said, "Look, there's
> the helix, and that damned woman just won't see it."
> He picked it up, of course.[19]

Considering the abnormality of the situation at King's, it is
difficult to blame Wilkins for failing to ask permission; surely
by then relations were too strained to allow him comfortably
to ask; and it is not uncommon, in any case, for the work that
is being done in any laboratory to be accessible to everyone
under the roof without formalities. But what had become of
Rosalind's report, written exactly a year earlier than the date of
this conversation with Watson, in which the "damned woman"
had interpreted the B-form diffraction pattern as suggesting "a
helical structure (which must be very closely packed) containing probably 2, 3 or 4 co-axial nucleic acid chains per helical unit, and having the phosphate groups near the outside"? [20]

If Wilkins was convinced by February 1953 that Rosalind
was correct in placing the bases in the center and the backbone on the outside of the structure, what was it that convinced him, but *also* left him believing that she thought the
structure nonhelical? Granted that Rosalind was not willing to
commit herself to the helical nature of the A form, and this
permits a little superficial confusion; but clearly it is the B

form which was being talked of exclusively by Wilkins to Watson, and in any case the X-ray pictures of the two are not confusable.

The puzzle is insoluble on the basis of present information. What should be noted, however, is that Wilkins has no need to blame himself too much for his confidences to Watson. For there happened to be another source for exactly the same information, and that too apparently fell into Watson's hands.

In December 1952, only two months before Watson paid his February visit to King's, the biophysics committee of the Medical Research Council held a meeting. The meeting was at King's College, for Randall was a member of the committee, and on that occasion was its host. At the meeting, Randall circulated a report covering recent work done in the laboratory under his direction—this was a matter of course—and the report included a concise summary, signed by Rosalind and Gosling, of the results of their X-ray studies of calf thymus DNA. Now, among those present was another committee member, Max Perutz, a very distinguished member of the Cavendish Laboratory at Cambridge. In the natural course of events, Perutz received a copy of the report. About this Perutz has written,

> As far as I can remember, Crick heard about the existence of the report from Wilkins, with whom he had frequent contact, and either he or Watson asked me if they could see it. I realized later that, as a matter of courtesy, I should have asked Randall for permission to show it to Watson and Crick, but in 1953 I was inexperienced and casual in administrative matters, and since the report was not confidential, I saw no reason for withholding it.[21]

Nor was there any reason for withholding it, not, at least, in any normal situation. Watson writes about this in a quite different vein. At this time he and Crick had had one of the great insights which led them to the structure of DNA, and he was very elated, as he had every reason to be, to be assured that what they had in mind was not incompatible with

the experimental data. By experimental data, he meant Rosalind's. As he admits, neither Rosalind nor anyone else at King's
had directly provided this information, and none of them realized what Crick and Watson had in their hands. The data had
been provided because Max Perutz was a member of the Medical Research Council's biophysics committee, which Watson
describes as "a committee appointed . . . to look into the research activities of Randall's lab. Since Randall wished to
convince the outside committee that he had a productive research group, he had instructed his people to draw up a
comprehensive summary of their accomplishments. . . . As
soon as Max saw the sections by Rosy and Maurice, he brought
the report . . . to Francis and me." [22]

There are several strands here to be sorted out. The suggestion that the committee was a special one set up to examine
the productivity of Randall's laboratory is, according to Perutz,
inaccurate. A committee set up in 1947 to " 'advise regarding
the scheme of research in biophysics under the direction of
Professor J. T. Randall' " was dissolved before the end of 1947.
The committee which existed in 1952 was formed " 'to advise
and assist the Medical Research Council in promoting research
work over the whole field of biophysics in relation to medicine.' " [23] What is significant is that once again Watson has
taken the opportunity to suggest that such chaos or inadequacy
prevailed at King's College that it had removed itself from
serious consideration as a place at which science was done.
Another strand is the exact part played by Max Perutz in the
transmission of the information. When *The Double Helix*
appeared, Perutz was understandably distressed by Watson's
description of his action. He has pointed out dryly that the report he showed to Crick and Watson contained, after all, no
data that "Watson had not already heard about from Miss
Franklin and Wilkins themselves. It did contain one important piece of crystallographic information useful to Crick;
however, Crick might have had this more than a year earlier
if Watson had taken notes at a seminar given by Miss Franklin." [24]

Though Watson subsequently denied, in a reply to Perutz's

published rebuke, that "the King's lab was generally open with all their data," [25] Perutz had a comment upon this too: "It is interesting that a drawing of the 'B' patterns from squid sperm * is also contained in a letter from Wilkins to Crick written before Christmas 1952. . . . This clearly shows that Wilkins disclosed many, even though perhaps not all, of the data obtained at King's to either Watson or Crick." [26]

Whether a laboratory is under any obligation to be generally open with data is an arguable point. Certainly there are circumstances in which reasonable prudence would advise it not to be. In this case it is evident that what was being done at King's was communicated to Cambridge in a variety of ways, some open and some at best accidental.

The question is, of course, whether any of this made a difference. The answer is plainly that it did. The difference was that of time. It appears that Crick and Watson had, on February 5, 1953, nothing in mind which permitted them *then* to build a conclusive model of DNA, but between February 6 and February 28, a successful model had become possible.[27] Most certainly the data received from King's were not *all* that was required to allow them to do this, but equally certainly they were essential. Not only did Rosalind's density data indicate the possibility of a two-chain model, but the diffraction pattern she obtained in the B-form photographs provided evidence of the diameter of the helix. More than this, "Rosy had hit it right in wanting the bases in the center and the backbone outside," [28] a configuration to which Watson had been, for his own part, rather resistant.[29]

None of this was information which Crick or Watson had developed through independent experimentation. No doubt it would have been accessible to them if they had tried. But such experimenting would have been, at the very least, time-consuming; and the more so in Watson's case, as he possessed at the time little experience in the methods by which it was done. This means, if nothing else, that the model based upon independently gathered experimental information could not have seen the light of day by the end of February 1953. Time is

---

* The work on squid sperm was done by Wilkins himself.

what counts in obtaining priority of discovery. Quite unknowingly, Rosalind gave to her rivals not only the gift of information, but of time.

She lost.

NINE

Winner Take All

It is always hard to argue with success, and to try it is idiocy. This is as true in science as anywhere; indeed, it is truer. Scientists value success as much as anyone, and success in scientific terms has a harder core to it than is usually possible in other fields, if only because it rests upon arguments which are immune to the whims of taste, and achievements which can never be entirely accounted for by luck. What is there that can be said, then, except that Watson and Crick published first, and published well and truly? They succeeded. The claim to priority of publication of the correct structure of DNA is theirs for all time, winner take all.

Nor is there ever much point in trying to make an arresting tale out of those who also ran. The best excuse for failure, however relative a failure, cannot help but sound feeble; "that's too bad" is about the most that sympathy can muster. To make excuses for Rosalind, who was the last person to choose to make them for herself, would be rather impertinent; one way and another, she has been subjected to enough impertinence, and I have no wish to add to it. But I should like, while pausing to praise Watson and Crick, to try to separate their enormous accomplishment, which was very real, from the faint air of rather mystical insight which has come to surround it, and for which there is small justification.

This is not to belittle what they did. The structure which they proposed for DNA is beautiful and true. It has a natural simplicity which convinces; it answers questions which have proved it from the beginning to be the uniquely correct answer to an immensely difficult puzzle; it has gone on answering enough more questions to make it the basis of a new field of science. Nothing has arisen in the course of subsequent work which has in any way contradicted it, and nothing is ever likely to, for all subsequent work has unfailingly confirmed and reconfirmed it.

That what Watson and Crick did in order to come up with their brilliant perception was to put together many fragments of pre-existing information is no argument against their accomplishment, for this is what science consists of and, in any case, no form of creativity should ever be assessed in terms of a plunge into the black waters of total ignorance in the hope of finding a possible oyster that may or may not contain a pearl. To rearrange a universe is creativity enough for anyone short of God, and this is very close to what they did, something within the capacity of few indeed.

This is enough, and more than enough. There is no need to suggest that it was also less a perception on the part of two scientists than a mystical revelation. How this notion crept in is clear; there is also some reason to guess that one motive which might have moved Watson to write *The Double Helix,* and confess that revelation had little to do with it, was a desire at last to put the notion to rest. But what had given rise to it from the outset was the paper that Watson and Crick published in *Nature* on April 25, 1953—"A Structure for Deoxyribose Nucleic Acid," [1]—which, for all the stunning content and the exquisite style, has been fairly characterized as being only a little longer than the Ten Commandments, and rather similar in its quality of appearing to have been handed down by God to the prophets waiting on Mount Sinai.

The temptation was tremendous. Probably every scientist in

the world has dreamed of publishing the perfect, and perfectly pure, perception, the paper so original that it cites no authorities, acknowledges no sources, has no history. There have been a few such, very few, and not many of these are modern. They all have the look of rather awesome genius, and command a rather awed admiration; and most of them, it might be said, are in mathematical or theoretical fields where data, in the sense of the accumulated products of experiment or research, are not essential components of perceptivity. Watson and Crick's *Nature* paper, less than a thousand words long, comes as close as is possible to suggesting pure perception, missing by very little. But quite apart from the natural temptation to present it this way, they were also, no doubt, somewhat inhibited by circumstances. What came their way from King's College, having come their way unorthodoxly, could not be credited in an orthodox way, and this explains a paragraph otherwise a little baffling, and perhaps somewhat elusive:

> The previously published X-ray data on deoxyribose nucleic acid are insufficient for a rigorous test of our structure.* So far as we can tell, it is roughly compatible with the experimental data, but it must be regarded as unproved until it has been checked against more exact results. Some of these are given in the following communications. We were not aware of the details of the results presented there when we devised our structure, which rests mainly though not entirely on published experimental data and stereochemical arguments.[2]

The "following communications" are two supporting papers which appeared in the same issue of *Nature*, April 25, 1953, with the Watson-Crick paper that proposed the DNA structure, all three appearing under the modest general heading, *Molecular Structure of Nucleic Acids*. These supporting pa-

---

* Here the following are cited: Astbury, W. T. Symp. Soc. Exp. Biol. 1, Nucleic Acids, 66 (Camb. Univ. Press, 1947). Wilkins, M. H. F., and Randall, J. T. *Biochim. et Biophys.* Acta *10*, 192 (1953).

pers were "Molecular Structure of Deoxypentose Nucleic Acids," signed by M. H. F. Wilkins, A. R. Stokes, and H. R. Wilson,[3] and "Molecular Configuration in Sodium Thymonucleate," by Rosalind E. Franklin and R. G. Gosling.[4] Supporting papers are what they are, offering experimental evidence which confirms the Watson-Crick structure; and doubtless they were not seen at Cambridge before the structure paper was written. What was known there, however, was a good deal of unpublished data accumulated at King's, and chiefly by Rosalind, which is dimly acknowledged, if acknowledgment can be read at all into the statement that "our structure . . . rests mainly though not entirely on published experimental data and stereochemical arguments."

Altogether, all things considered, Crick and Watson may not have had much choice, and what choice there was—which we will come to in time—was not one that pleased them. But here lies the beginning of the notion that Rosalind was not only beaten by a mile, but in terms of time, by years and years; and that her DNA work was at best irrelevant, a blundering failure. And this is not in the least true.

What Crick and Watson did was to put together in a way which no one had seen before several isolated pieces of information about DNA. One of the most important of these was provided by Erwin Chargaff, a contribution which Watson assesses in *The Double Helix*. What Chargaff and his students had done, Watson says, was to analyze various DNA samples for the relative proportions of their purine and pyrimidine bases, and to discover that in all their DNA preparations the number of adenine (A) molecules was very similar to the number of thymine (T) molecules, while the number of guanine (G) molecules was correspondingly close to the number of cytosine (C) molecules. This was true although the proportions of adenine and thymine groups varied with the biological origin of the material. In some organisms, the DNA produced had a higher proportion of A and T, in others, of G and C. Chargaff did not offer an explanation for his results, and no explanation immediately leaped to anyone's mind.[5]

Watson read Chargaff's work, and reported its contents to

Crick, though for the time being, the possible significance of it did not strike them. According to Robert Olby's chronology of the events leading up to the discovery of the structure of DNA, it was in June 1952 that Crick, with some inspiration from a young mathematician, John Griffiths, realized, "Why, my God, if you have complementary pairing you are bound to get a 1 : 1 ratio," but the further implications seem not to have been picked up until February 20, 1953.[6] Then Jerry Donohue, an American crystallographer visiting for a time at the Cavendish, provided some useful advice and criticism. Donohue's special field of knowledge lay in hydrogen bonds, about which, according to Watson, "next to Linus himself, Jerry knew more . . . than anyone else in the world."[7] By shifting, on Donohue's advice, from the enol to the keto forms of the bases, Watson took an essential step toward the correct structure, but these forms fitted even less well than the previously hypothesized ones into the sugar-phosphate backbone which Watson was still inclined to locate in the center of the molecule.

It was when this new information was put together with Rosalind's well-defended proposal that the backbone lay on the *outside* that great things happened. Olby puts it this way,

> Evidence *so far collected* * suggests that this successful attempt in 1953 to determine the structure of DNA took from Friday, February 6, when Watson took Pauling's DNA manuscript [8] with him to King's College, London, until Saturday, February 28, when Crick retired to bed exhausted after nearly a week of model building. At King's, Watson learned from Wilkins that the density data did not after all rule out two-chain models, and that the sugar-phosphate chains must, as Franklin had stated in Watson's presence in 1951, be on the outside. At the end of the first week back in Cambridge, Watson had come round to this view. Crick recalls an earlier incident when Watson was complaining of the difficulties he was having in the attempt to intertwine the back-

* Olby's italics.

bones on the inside of the structure. Crick's reaction was to ask, "Then why don't you put them on the outside?" Watson replied: "That would be too easy," which prompted the retort: "Then why don't you do it!" Watson himself wrote: "Finally . . . I admitted that my reluctance to place the bases inside partially arose from the suspicion that it would be possible to build an almost infinite number of models of this type. Then we would have the impossible task of deciding which one was right. But the real stumbling block was the bases. As long as they were outside, we did not have to consider them. If they were pushed inside, the frightful problem existed of how to pack together two or more chains with irregular sequence of bases. Here Francis had to admit that he saw not the slightest ray of light." [9]

From this, two things at least are evident. One is that model building, as an approach to structure determination, is not limitless in its powers. Unless it incorporates a great deal of defining information, it can easily produce an almost endless number of unprovable and untestable hypotheses. Quite evidently Crick and Watson had come so far but no further. From this the second point follows. Without the information acquired by Watson on his visit to King's, the successful model of DNA would not have been completed at Cambridge by February 28, or indeed, any date close to that.

For this contribution of time, Watson and Crick offered thanks in their paper. "We have also," they said, "been stimulated by a knowledge of the general nature of the unpublished experimental results and ideas of Dr. M. H. F. Wilkins, Dr. R. E. Franklin, and their co-workers at King's College, London." [10]

"Stimulated" seems a minimal word for it.

On February 28, then, Crick and Watson completed their model. They were charmed with it, as well they should have been. By every test they could apply, it "worked." It did more than that. As they put it, in the most delightful throwaway line ever incorporated into a scientific paper, "It has not escaped our notice that the specific pairing we have postulated immediately suggests a possible copying mechanism for the genetic material." [11] Their structure *explained*.

There was, of course, no need for them to hesitate about publishing this beautiful thing. What there might have been some hesitation about was the form the publication should take. An offer might have been made, in view of everything, to the people at King's for joint publication, but that this possibility existed seems to have crossed no one's mind. But that Wilkins, at least, was not immediately informed seems perhaps a little surprising. That Watson recalls having mused over the fact that Maurice must soon be told indicates that he, if no one else, caused some conscientious qualms.[12] Letters promptly went out to places further away than London, which can be reached from Cambridge by telephone quite inexpensively, but Wilkins apparently was left to learn of the triumphant joy at the Cavendish until the paper was written. Olby says,

On that Wednesday [March 18] Wilkins received the Watson-Crick typescript, or perhaps several versions of it, for on the 18th he wrote:

"Thanks for the MS. I was a bit peeved because I was convinced that the 1 : 1 purine-pyrimidine ratio was significant . . ." and adds, "Just heard this moment of a new entrant in the helical rat-race. It seems that they (R.F. and G.) should publish something too. (They have it all written.)"

The only construction that I can put on Wilkins' letter is that the first he knew of the structure was when he got these MSS. Yet Watson writes in his book that when the model had been completed he told Francis he would write that afternoon to Luria and Delbrück: "It was also arranged that John Kendrew would call up Maurice to say he should come out to see what Francis and I had devised." Yet Watson's letter to Delbrück is dated March 12th. Franklin and Gosling wrote their draft on or before the 17th,

and Wilkins first learnt about the double helix on
March 18th.[13]

No doubt elation, or embarrassment, or some combination
of the two, might explain why Wilkins heard about the triumph of his Cambridge friends only just in time to be able to
contribute a supporting paper to the same issue of *Nature* in
which the triumph was announced. The question of joint publication never arose. Nor, of course, did it arise with respect to
Rosalind.

That she and Gosling had a paper all written in advance of
March 17 is in itself interesting. Clearly her own work had
reached a point, quite independently of anything that might
be going on at Cambridge, where she felt that something
worth publishing had been achieved. It has been natural to
assume that the paper by her and Gosling which appeared in
the April 25 issue of *Nature*—"Molecular configuration in sodium Thymonucleate"—is the same as the one prepared before the news from Cambridge arrived. This paper discusses
both the A and B forms of DNA—including the X-ray photograph of the B form with its helical cross—and, in addition to
affirming in published form the location of the phosphate
groups on the outside of the molecule, opts for a two-chain
helix in the B form, and probably in the A form as well.

In 1969, Aaron Klug discovered among the papers Rosalind
left at her death a manuscript in draft, dated March 17, 1953,
which clearly is what Wilkins was referring to as the paper
"all written," and which, though a precursor of the *Nature* paper, is not identical to it. This also proposes that the B form
is helical, that the diameter of the helix is about 20 Å, and
that "a [crystallographically] *very important part of the molecule lies on a helix of this diameter.* This can only be the phosphate groups. . . . Thus, if the structure is helical, we find
that the phosphate groups lie on a helix of a diameter about
20 Å, and the sugar and base groups must accordingly be
turned inwards towards the helical axis." [14] And from there
she proposes two equally spaced coaxial helices.

She came very much closer to the discovery of the double

helix than she has usually been credited with doing, and the date on the draft paper makes quite clear when she did so. The existence of this paper does not challenge for a moment the priority of Crick and Watson's structure; but it does make nonsense of a great deal which has gotten into print, some of it by Watson. The discovery of the draft paper was communicated by Klug to Watson, who responded with amazement, suggesting that perhaps Rosalind had been inspired by events in Cambridge, passed along to her by Wilkins. But such inspiration was not necessary. Klug had previously studied Rosalind's notebooks preparatory to writing an article about her contribution to the DNA discovery,[15] and was able to confirm from them the direction in which her thinking was moving well before she wrote the draft paper of March 17.

She was close; she was very close; the pieces were beginning to come together in the right way, and it should be remembered that when the pieces begin to fall in place, they often fall quite fast. It should also be emphasized that Rosalind was working essentially alone. The contrast between what was available to Crick and Watson in Cambridge and what Rosalind had for stimulation and support at King's is a staggering one. The perception of the significance of Chargaff's work in providing the 1 : 1 ratio was a splendid one, but it was assisted by a useful conversation with a mathematician. How long would Watson have wrestled with the problem of the bases had Donohue not been present to suggest the keto instead of the enol forms? Watson had Crick, and Crick had Watson. Rosalind's available source for conversation, advice, helpful speculation was Luzzati, in Paris. Because she had done all her previous crystallographic work in France, she was not even acquainted at the time with the people in England who were working in the field. At King's she had Gosling to talk to, and—to a limited degree—Randall. In view of all this, the pace at which she was moving was, indeed, comparable to that of Watson and Crick, and to see it so is not a case of special pleading.

One indication of her nearness to the solution of the problem was her instant appreciation of the rightness of the Watson-Crick model. She amazed Watson, or so he says, by producing no irrelevancies to cast doubt upon the correctness of the double helix. His fears about the failings of "her sharp, stubborn mind, caught in her self-made antihelical trap" [16] proved to be illusory, very likely because the antihelical trap was of Watson's devising, though he prefers to think it was because X-ray evidence itself had begun "forcing her more than she cared to admit" [17] in the direction of accepting a helical form. The meeting with Rosalind at which the Cambridge model was displayed to the King's College group turned out not to be as dreadful as Watson had anticipated, but he had had warning of a change in her attitude from a visit paid to King's by Crick, at which Rosalind appeared ready to "exchange unconcealed hostility for conversation between equals." [18] When Rosalind demonstrated to Crick how "foolproof was her assertion that the sugar-phosphate backbone was on the outside of the molecule," Watson subsequently reflected upon this in a curious vein: "Her past uncompromising statements on this matter thus reflected first-rate science, not the outpourings of a misguided feminist." [19]

The notion that accurate statements made by a woman scientist are first to be regarded as likely outpourings of feminism, and only under the strong pressure of irrefutable demonstration as science is Watson's own contribution. So is the fictional "Rosy." Unfortunately these Watsonisms have been widely believed. From scientific publications to discussions in the popular press,[20] both Rosalind's views and her contributions to the solution of the structure of DNA, and sometimes her personality as well, have been either misrepresented or overlooked, and in all cases the source for this treatment appears to have been Watson.

Yet the facts are not all that difficult to ascertain or to grasp. If one thinks of the structure of DNA as a prize awarded at the end of a race, then Rosalind lost the race; by how much is debatable, but by rather less of a margin, it is safe to say, than the one which would have occurred in the opposite direction if Crick and Watson had been required to develop from their own experimentation the data provided, never mind how, from

King's.[21] Her contribution to the successful pair who proposed the successful structure was, as Klug has put it, "crucial." And he is worth quoting at length:

> She discovered the B form, recognized that two states of the DNA molecule existed and defined conditions for the transition. From early on she realized that any correct model must have the phosphate groups on the outside of the molecule. She laid the basis for the quantitative study of the diffraction patterns, and after the formulation of the Watson-Crick model she demonstrated that a double helix was consistent with the X-ray patterns of both the A and B forms . . .
>
> . . . if for a time Franklin was moving in the wrong direction in one aspect . . . then there are clear indications that equally she was moving correctly in another. In the first paper [22] Franklin also gave attention to the problem of the packing of the bases. She discussed the existence of small stable aggregates of molecules linked by hydrogen bonds between their base groups and with their phosphate groups exposed to the aqueous medium. She discusses the obvious difficulty of packing a sequence of bases which follow no particular crystallographic order and the state of her thinking can be seen in the following extract from her March 1953 paper:
>
> "On the other hand it also seems improbable that purine and pyrimidine groups, which differ from one another considerably in shape and size, could be interchangeable in a structure as highly ordered as structure A. A possible solution, therefore, is that in structure A cytosine and thymine are interchangeable, and adenine and quanine are interchangeable, while a purine and a pyrimidine are not. . . . In this way an infinite variety of nucleotide sequences would be possible, to explain the biological specificity of DNA."

Base interchangeability is, of course, a long way from the final truth of base pairing, but in the context of the crystallographic analyses in which Franklin was engaged—an analysis which could provide a solution to the regularly repeating parts of the structure—the idea would have been essential to fitting in the variable parts.[23]

The vital information which was leading Rosalind toward the correct solution, she had developed for herself, experimentally, and all of it was available to Watson and Crick. From them she had nothing. If the image of a race is still to be used, it should be observed that it was run under handicappers' rules, with the advantage assigned not to her, but to them.

It is also worth noting that Rosalind herself had no idea, ever, how crucial her contributions were to the Watson-Crick structure, because she never knew the extent to which her data had passed into their hands or had been talked over with Wilkins. All that she was aware of having provided was what went into her seminar in November 1951, and what she had to say in criticism of the early Watson-Crick model. It is rather interesting to reflect that, if Watson had understood what Rosalind had been saying in that seminar—had taken notes, remembered correctly, or asked modestly for an explanation—he might have brought back to Crick information which would have enabled them, had they put it all together as brilliantly as they later did, to solve the problem much sooner.

But then, it is not really feasible to come home from a lecture, utilize the new learning, and produce a fresh and insightful perception without acknowledging the source of inspiration. In order to make exclusive claims, it is necessary to gather information more indirectly; and as Watson frankly admitted in *The Double Helix*, this is exactly what he did. There are matters raised by this method of proceeding that offer endless opportunity for scrutiny and debate. They are worthy of scrutiny and debate. The opportunity has been lost in the clouds of glory which surround the Watson-Crick triumph, and the moral of the story, as it has unremittingly been presented, has been a crude and simple one. Winner take all.

TEN

"What She Touched, She Adorned" [1]

During the first two weeks of March, Rosalind worked on the paper that, in her mind, was her farewell both to King's College and to the DNA problem. She was in the process of removing herself to Bernal's lab at Birkbeck College, and she had learned that the DNA problem was not to go with her. Randall had made that perfectly clear. She could leave if she liked, she could take her Turner-Newall Fellowship with her, but there was to be no mistake or misunderstanding about the portability of DNA, which remained the exclusive property of King's. She did not dispute this so far as material and data were concerned; when admonished sternly "to stop thinking about DNA entirely," she was puzzled. [2] But those were the conditions, and Rosalind had not much choice but to accept them.

It cannot have been easy for her to abandon a problem which had occupied her for two years, and which she knew she had brought a long way, just short of the whole way, toward a successful conclusion. Toward the end of 1952, she had begun to lose interest in the A form, which had turned out to be, as far as she was concerned, far less promising than she had hoped, and had turned back to considering the possibilities of the B form which—to judge by the entries in her notebooks and the draft paper of March 17—was leading her in the direction which proved to be the right one, and which certainly she found very provocative.* It was so unlike Rosalind to have surrendered, to have abandoned a problem that she had every reason to believe was not only soluble, but not very far from being solved, that it must be assumed that the pressures upon her to leave King's were very strong ones, strong enough to overcome her tireless persistence—or her native stubbornness, whichever one cares to call it.

Whether she was disappointed when news of the Crick-Watson triumph at Cambridge finally filtered through to King's no one knows. If she was, she gave no indication of it, and her disappointment cannot have been very acute. She sat down to redraft her paper for *Nature* in the new form of a supporting document for the Watson-Crick structure with every sign of pleasure, none of bitterness or chagrin. This is exactly what one might have expected.

To begin with, the structure they proposed genuinely delighted her, as it did everyone capable of grasping its simple beauty. Watson seems to misunderstand what she felt, and to assume that her conversion rested upon a long-delayed appreciation of model building as a serious approach to science. To assume, as he does, that Rosalind regarded it as an "easy resort of slackers who wanted to avoid the hard work necessitated by an honest scientific career" [4] is to be quite inventive. Plainly he did not trouble to observe that Rosalind had done extensive model building in her graphite work and was, therefore, unlikely to think of it as unserious, ineffective, or the lazy resort of idle minds. There was really nothing for Rosalind to

---

* Rosalind was, for example, convinced by the time she wrote her March 17 draft that the B form was helical, with two coaxial chains. How poorly she and Wilkins communicated is indicated by his being able to write to Aaron Klug in 1969, and with evident sincerity, that though Klug might well be right that Rosalind fairly consistently thought the B form was helical, she kept this view to herself. [3]

be converted to. The elegance, the rightness, of the Crick-Watson structure she was in a better position to appreciate than anyone else, having approached it so closely herself. The techniques they used in order to arrive at it were ones concerning which she cherished no prejudices. Nothing was ever as complicated as Watson thinks.

This seems never to have prevented him from finding complications anyway. When at last he found it in his heart to offer Rosalind a little mild sympathy, it took a curious form. Her difficulties, he concludes, sprang from her rebellious resentment at receiving from King's no "formal recognition" of her ability.[5] This can only mean, if it means anything, that Rosalind would have liked a title on her door, and resented the lack of it. But from the day when she left CURA to the end of her career she always lacked "formal" recognition in that sense. She never held a professorship, never launched herself on the upward climb through the academic ranks, never "ran" anything but her own work. She avoided all these ornaments to a career, refusing, for example, to join the teaching side of Birkbeck, though that was where the power and the potential for "formal" glory lay.

To no one could it have mattered less than it did to Rosalind that she was outside the hierarchy at King's. She did object to the frigidity of the internal climate, and certainly she disliked wasting her breath in arguments which failed to communicate anything much at all to those she was arguing with. If Wilkins really did not know that Rosalind had proposed in 1951 that the B form was helical, and by early 1953 was convinced of it, or that she had evidence upon which to base her reasoning, then something was very wrong indeed at King's, and Rosalind had good cause to despair. She liked problems to be solved, that was what she was a scientist for. She wanted the DNA problem to be solved, and if what she would have liked best was to solve it herself—she was not unambitious in the least —this mattered less than that someone, somewhere, would come up with the right answer.

It was no secret to her that what Wilkins had in mind to do as soon as she left King's was to repeat her work; more than two years of productive effort was about to be overlooked; the solution of the problem, at least as far as King's was concerned, was to be pushed back in time; and none of this seemed sensible to Rosalind, which is not surprising. It is natural enough that when the problem was solved, she was delighted. And it was beautifully solved. Rosalind required no persuasion whatever to accept the Watson-Crick structure. She took one long look, and was convinced, simply because she was a very good scientist who knew her subject and could not help but recognize the perfect solution when she saw it. It was, come to think of it, one of the best compliments Crick and Watson ever received.

That their smashing success converted much of what was in Rosalind's draft paper into secondary material, that her discoveries were abruptly swallowed up by their larger discovery, seems not to have troubled her at all; or if it did, nothing she wrote or said testified to it. The paper which had represented, while she and Gosling were working on it in the first weeks of March, a new high-water mark in research into the structure of DNA, was promptly redrafted into a supporting paper. It reads serenely in the second version, as one might expect; there is no hint in it of "yes, but I saw that first."

That her work had contributed very significantly to Crick and Watson's structure—and that this might be said to more than justify it—Rosalind never quite realized, and for the simple reason that she had no notion at all that anyone outside King's had access to her unpublished results, much less that anyone had used them. This is, of course, very curious; but there is no reason to think that the situation was otherwise. Rosalind knew only that Watson had come to her seminar in the fall of 1951, that he had listened rather opaquely to her early results, and that he had not subsequently indicated the slightest interest in them. She knew that in the course of two years, he had heard what she had to say without showing any enthusiasm for it, and that even when she was right, his reaction was rather negative. As he puts it himself, commenting upon one of the occasions when she was right, her objections were not merely perverse.

Not in circumstances such as these does one believe one's viewpoints to be valued. There was no way in which Rosalind could know what Watson asked of Wilkins, or what Wilkins replied. She seems to have taken the Cambridge structure as it was presented, as a work of perception, insight, and inspiration, and though she was pleased that it confirmed her work precisely as her work confirmed it, she did not know that, indeed, it incorporated her work. Whether she would have been pleased by the use to which her findings had been put, or resentful at both the way in which they were obtained and the way in which they were left unacknowledged, is a nice question to speculate about. My own guess—freely disputable—is that Rosalind might well have risen like a goddess in her wrath, and that the thunderbolts might have been memorable.

But she did not know. She looked at the Crick-Watson structure, and she vastly admired it. Her admiration for Crick in particular was profound. She regarded him as a genius, a scientist of unique ability, and whenever she spoke of him to me, it was in these terms. She also liked him very much, a feeling which she never quite extended to Watson, whom she respected, but only across a certain cool distance.

Rosalind's departure from King's was made graceless, though not by her doing. The prohibition that forbade her to work with DNA, or even to give it so much as a thought, may have appeared rather meaningless in the wake of the Cambridge triumph, but it was not rescinded. This affected Rosalind not at all. Her interest, start to finish, had been in the DNA structure; if Wilkins chose, as in fact he did, to go on collecting and refining a mass of further information, he was welcome to the job, for Rosalind's curiosity was unprovoked. The person who was affected was Ray Gosling, who had his doctoral thesis well under way, and found himself abruptly deprived of a supervisor. The word came down: Rosalind and Gosling were to have no contact whatever. They were not to meet, they were not to discuss orally or in writing anything whatever, at least not until such time as Gosling had his degree, and had passed out of the care, and the control, of King's College. This decree struck Rosalind as fussy and tiresome—"just the sort of thing they do there" [6]—and even more than this, as funny.

She was attached to Gosling, with whom she had got on very well, and through difficult times. They had nothing very evident in common; Gosling was tall, athletic, genial, outgoing, mild of temperament, and there were those who doubted strongly whether he would find Rosalind's seriousness, and her intolerance of any lack of seriousness, as endurable. There appears never to have been serious friction. Gosling, for all his amiability, had no objection to Rosalind's liking for hard argument, and if she worked him hard, he was remarkably uncomplaining about it. On her part, she appreciated his loyalty and liked his friendliness. To let him down was unthinkable; and so they met, communicated, discussed, and amused themselves tremendously by pretending that it was all clandestine. Not only Gosling's thesis, but several joint publications emerged from this mildly secret collaboration, for Rosalind, the last of DNA.[7]

This mattered not at all. She had a more difficult problem in mind. She did not settle down to it at once; before the famous April 25 issue of *Nature* had appeared, Rosalind was off to a meeting in Aachen, and in June she was in Paris for a colloquium and a reunion. At the first she read a paper on "The Mechanism of Crystallite Growth in Carbons," and at the second, one on "Le rôle de l'eau dans l'acide graphitique," [8] both of them retrospective, a backward glance at the carbon work which she no longer pursued actively, but which she never quite abandoned.*

At the end of the summer, she paid her first visit to the United States, to attend another carbon conference; she spent a few days in Philadelphia with David and me, and as far as I can remember, had more to say about a recent trip to Israel where she had done some fairly adventurous hitchhiking around the Dead Sea than about DNA. She was, anyway, already ab-

---

* Her last publication on carbons, "Homogenous and heterogenous graphitization of carbon," appeared in *Nature* in 1956.

sorbed in her prospects for Birkbeck, and she was full of optimism. No discouraging experience ever seemed to discourage Rosalind for long.

On the whole, her optimism was justified. J. D. Bernal, who headed the Birkbeck lab, was a man whom Rosalind could admire as a scientist; on the political side, his views failed to impress her. Bernal was a long-time and devoted Marxist, concerning which Rosalind was calm. Her own political attitudes were conventional, perfunctory, and unless she was in a teasing mood and willing to bait the opposition, unargumentative. On her own, she canvassed once or twice for a local Labour candidate, not enjoying it very much; to put it in American terms, this is roughly equivalent to saying that she addressed envelopes on behalf of a Democrat running for Congress.

The political atmosphere at Birkbeck occasionally annoyed her; she wrote me in December 1953, "Birkbeck is an improvement over King's, as it couldn't fail to be. But the disadvantages of Bernal's group are obvious—a lot of narrow-mindedness, and obstruction directed especially against those who are not Party members." [9] More often, I think, she was amused; in March 1957, she wrote, "Reactions in Birkbeck to Hungarian things were quite amusing—really all the Party members were quite shaken, though I don't know any who left the Party." [10] In October she was still objective, and still fundamentally indifferent: "Bernal was somewhat shaken by Hungary, but I think only temporarily. On the rare occasions when I heard him speak about it, he got emotional and confused, and a few months later he was again talking in the old Party clichés." [11]

She and Bernal were, in fact, never particularly close; but she may have benefited all the same from his convictions, for he held to the rather old-fashioned Communist notions about the equality of male and female workers, and was well known for his willingness to accept women students, to encourage them, to promote their careers, to find opportunities for them. That Rosalind took all this largely for granted, seeing it much more as the natural attribute of the highly intelligent man rather than of the political one, is merely reflective of her own attitudes. But they got on well enough; he was a very superior scientist, and that was sufficient for Rosalind.

He also had a problem on hand. As early as 1935, he had begun crystallographic work on the structure of tobacco mosaic virus—TMV—in an effort so far ahead of its time that the practical lengths to which it could be carried were rather severely limited.* This problem Rosalind picked up where he had left it; she wrote in December 1953, "It's been slow starting up here, but I still think it might work out all right in the end. I'm starting X-ray work on viruses (the old TMV to begin with.)" [12] The laconic understatement is very typical of Rosalind.

For the rest of her life, she worked on viruses, at Birkbeck. She was peculiarly accommodated, in an old house in Torrington Square which had been severely battered by bombs during the war, and afterward cobbled together rather inadequately —it has since been torn down to make room for Birkbeck's new buildings. I visited her there one rainy afternoon a few years later, and was surprised to find that the roof leaked. Rosalind had beakers and pots neatly arranged to catch the drips. When Katarina Kranjc, the Yugoslavian crystallographer who had become one of Rosalind's friends, paid a call, she was shocked,

> Do you know in what conditions she worked? Her tiny room was on the fifth floor, and the X-ray equipment in the cellar. How many times a day did she have to climb those stairs? There was no lift, and the last of the stairs was twisting. I was amazed, and asked why that room was not assigned to a younger person, and why such a prominent scientist must work in the worst conditions. Rosalind did not complain, and refused to discuss it; she said she was content with it as it was. [13]

This was very probably true. The trappings of success—or prominence—were never what Rosalind was looking for. She had a great deal of natural, and perfectly genuine, simplicity.

---

* In this early work on TMV, Bernal collaborated with Isidor Fankuchen.

Her apartment in Drayton Gardens was comfortable and attractive, well-furnished and beautifully housekept; she liked nice surroundings, but luxury bored her. She had an independent income which she drew on infrequently and reluctantly, preferring to live on what she earned, which was adequate but modest.

She was an expert on cheap travel, the kind she most enjoyed; that on one occasion she went to Israel traveling steerage on a slow boat seemed so curious to some of the people she worked with at Birkbeck that they assumed her journey was a kind of pilgrimage, accomplished as uncomfortably as possible out of some obscure but dedicated motives. But Rosalind was not in the least given to making pilgrimages anywhere. She enjoyed being cheerful in discomfort, and she once confided that she liked to travel with very little money "because then you need your wits," [14] which reminds one strongly of those indomitable English ladies who sallied out to explore unlikely corners of the earth all through the nineteenth century, preferably on mule back or in a camel saddle and, no doubt, often dangerously.

I scoured through secondhand bookshops until I found a copy of Miss Irby and Miss Muir Mackenzie's *Travels in the Slavonic Provinces of Turkey-in-Europe* (with an introduction by Gladstone) to give to Rosalind; it was very nearly the only book on whose merits we agreed, though it made her regret that by the middle of the twentieth century "everything had got so easy, though in this case it's better for the Yugoslavs, I suppose." [15] She never really changed her mind about the desirability of traveling third class and in rough conditions; in 1957 she wrote of a holiday in Italy, "This was my first continental holiday by car . . . and I confirmed my impression that cars are undesirable. . . . Travelling around in a little tin box isolates one from the people and the atmosphere of the place in a way that I have never experienced before. I found myself eyeing with envy all rucksacks and tents." [16]

Anyone with this point of view is unlikely to object to a little discomfort in working conditions, even when it meant a leaking roof in every rainstorm. And besides, she was content. There were occasional difficulties; in the autumn of 1955 the Agricultural Research Council, which had been supporting some of her work, chose to withdraw its support, and Rosalind had a rousing fight on the subject with Sir William Slater, the ARC's director—and lost.* In time the loss was recovered; Rosalind wrote, "Did I tell you that we had applied to the U.S. Public Health Service for funds, having failed to get anything from anybody over here? Anyway, the grant has come through, and we are fixed up for the next 3 years, and are extremely grateful to all you kind American tax-payers." [17] But she got on well enough with everyone at Birkbeck. There were none of the serious storms with which she had been surrounded at King's.[18] And sometimes she got on better than well enough. In 1954 she was joined by Aaron Klug, a young South African; their collaboration was to be the longest, and the happiest, in Rosalind's working experience. It lasted, and lasted serenely, for the rest of her life.

Klug had the strength both of intellect and personality to argue back—exactly what Rosalind, who had no great opinion of meekness, always hoped for. He also had very great ability, and they appreciated each other; certainly she impressed him from the first. In his early days, he watched Rosalind working to produce suitable specimens of TMV for X-ray diffraction study, learning as she did so that TMV is among the world's more tricky, exasperating, and ungrateful substances for any such intention, and he saw immediately that if anyone could make it behave, that was Rosalind. "She *noticed*," he says. "She noticed everything. The fact that she produced the best specimens of TMV wasn't due to chance, or simple mechanical skills. It's an art, doing this, it's a matter of the pains she took, the way she nursed it, the keeping track of things, the

---

* Aaron Klug has said that the only time he ever saw Rosalind in tears was after her interview with Slater. Rosalind told me that she firmly believed she was denied funds "because the ARC refuses to support any project that has a woman directing it."

noticing. That's how discoveries are made. And this was one of Rosalind's greatest gifts." [19]

She was, indeed, one of the world's great experimental scientists. This is no mean achievement in itself. It may also be true that it was in her experimental work that the imagination and intuition she was in general rather cautious about using as a basis for reasoning found their natural expression. If she had been no more than methodical and painstaking her experimental work would never have risen to the remarkable levels that constantly characterized it; but she also had both a great store of knowledge to which she had unusually unconfused access, and that mysterious instinct, "feel," for experimentation. This can be seen as far back as her CURA work on coals; it was the determining ability that allowed her to produce the B form of DNA; it is evident everywhere in her work on viruses. This in itself was enough to make her an important scientist, but she added to it a close-to-equivalent "feel" for theory. This is not to say that she was one of the great, original, inventive theoreticians of her time, for this she was not; but her use of theory was almost unfailingly distinguished. At no point was she the experimenter who stopped at that, producing pretty data for other people to interpret, or finding the gap between the concrete and the theoretical an unbridgeable one. As Klug has put it, "she could marry the two. She just *moved,* she just *knew.*" [20]

In the last years of her life, Rosalind did beautiful work. How much Rosalind did on viruses between 1953 and 1958 is amazing. The list of publications is staggering: there were seventeen of them.[21] If a production rate of only somewhat over five publications a year seems small, then compare it with what other scientists manage to put out, and reconsider. Moreover, all this was pioneering work, the pace of which was relatively slow, and it was immensely complex work, far more testing, in a technical sense, than the DNA work. And, from the autumn of 1956 to the spring of 1958, Rosalind's health was failing; she had long interruptions when she was ill, or recuperating; she was often in pain and often close to exhaustion.

In spite of all this, the amount she accomplished on TMV is enormous.

It was also of very great importance. Tobacco mosaic virus has a remote and agricultural sound to it; but then, *Drosophila* is a fly, and its connection with genetic inheritance is not immediately obvious either. From the viewpoint of virus research, TMV is the classical virus, the first to be recognized, the first to be purified, the first to be studied by X-ray diffraction, the first to have its chemical progression worked out, the first to have its mechanism of regulation worked out, the first to be mutated in a test tube. From TMV most of the fundamental information about viruses was first obtained; then after it had begun to appear that its value as a source of knowledge was close to exhausted, TMV re-emerged into the limelight as the first substance in which regulatory protein-nucleic acid interaction was studied. Forty years after Bernal and Fankuchen took its X-ray photograph, it is still being investigated; in 1953, when Rosalind approached it, much of it was still mysterious. To begin with, very little in general was known about the structure of viruses, so that what she had accomplished by 1958 in working out a plan for the structure of TMV represented a giant step taken in something close to the dark. She did not produce then, or ever, a detailed structure, for this did not emerge until twelve years or more after her death. But when the details were there, their relationship to what Rosalind had discovered was plain.

Rather coincidentally, Watson had also done some work on TMV at Cambridge during one of the periods when Bragg's disapproval of the pursuit of DNA had required him to seek an alternative problem. In his comments upon his own work in *The Double Helix,* Watson seems satisfied with his accomplishments as a novice at X-ray diffraction work, though rather bored at the thought of unraveling the detailed structure of TMV. He did not, in fact, take it far. It is unfortunate that while seeking the significant structure factor which would prove TMV also to be helical, he hit upon the wrong reflections. But he was right in his guess. Rosalind, working with Kenneth Holmes, a graduate student, later found on the equator of the pattern the diagnostic reflection that showed TMV to be a helical structure, and later was able to fix the helical parameters.*

Behind many of the ideas on macromolecular structure which began to develop after the 1950s, Rosalind's discoveries may be found pointing the way. In the epilogue of *The Double Helix*, Watson pays his tribute, and he is not ungenerous, pointing out that what she did at King's has been increasingly appreciated as "superb," justly awarding her full credit for her contributions, and indicating his admiration of her tobacco mosaic virus work as well.[22] After Rosalind's death, Bernal wrote for *Nature* on the subject of her TMV work,

> Watson had put forth the hypothesis that the virus structure was . . . spiral, but one of a different order from that which existed in proteins and in deoxyribonucleic acid. Miss Franklin, with the help of very much better X-ray photographs than had hitherto been obtained, was able in essence to verify this hypothesis and to correct it in detail. It was at this point that the extremely fruitful cooperation began between Miss Franklin's unit and Fraenkel-Conrat at Berkeley, Caspar at Yale, and Schramm at Tübingen. Using the method of isomorphous replacement, she showed that the virus particle was not solid, as had previously been thought, but actually a hollow tube. . . . The combined methods of chemical preparation and X-ray examination in the hands of Miss Franklin and her associates was a valuable, and indeed a decisive, weapon in the analysis of these complex structures.

> As a scientist Miss Franklin was distinguished by extreme clarity and perfection in everything she un-

---

* This was done by the "heavy atom" or isomorphous replacement technique, where a mercury derivative of TMV was used, in order that the location of known and measured atoms in the diffraction pattern might be used to fix and measure the position of other atoms. After Rosalind's death, Holmes carried this work much further, with great distinction.

dertook. Her photographs are among the most beautiful X-ray photographs of any substance ever taken.
. . . She did nearly all this work with her own hands.
At the same time she proved to be an admirable director of a research team and inspired those who worked
with her to reach the same high standard." [23]

There is no one to deny any of it. But Aaron Klug said it best.
What she touched, she adorned.

ELEVEN

The Last Chapter

There was so little time. There were three untroubled years, a little more than that, but not quite three and a half. They were splendid years for Rosalind, when everything that she put her hand to went right, and when she had every reason to believe that in the future everything would go better still. The beguiling superstition that scientists invariably do their best work when young, and afterward—victims of some new form of gravity that operates only upon selected intellects—slide down the ladder rung by rung was not one that Rosalind believed in, and like other superstitions, this one may require credulousness in order to work.

Some scientists—like some pianists, painters, singers, writers, evangelists, and what have you—reach an early peak, some reach a late one, and some never reach anything that can be fairly described as a peak at all. Rosalind's abilities improved with time. Perhaps this was because there were ways in which she grew up slowly. Certainly she was in her thirties before the tensions which are natural to an intellectually precocious but emotionally uncertain schoolgirl wholly disappeared. When this happened, she lost nothing, and gained a kind of serene self-confidence that predicted everything good. When I saw her in New York in the summer of 1956, she seemed to me less defended, more open, more accessible than

I had ever seen her before. The bubble and sparkle were less restrained, more steadily present; and besides, she had become tremendously attractive. She was never a pretty girl, but she had been a handsome one who had flashing moments of stunning beauty; in her middle thirties, the beauty, and the animation which went with it, had emerged, were less intermittent, much less suppressed.

She was not in the least heavy or stiff. Indeed, she was the opposite. Certainly there was nothing stiff about Rosalind's delight in the kittens which our cat had recently produced. She was down on the floor playing with them almost as soon as she came in the door, and was—she said—"honored" that the mother coopted Rosalind as a nursery maid and nightly stowed the kittens in Rosalind's bed. But Rosalind had always liked cats very nearly as much as David and I did; what was much more of a change was that she liked parties, even large ones of the sort which had previously made her rather shy. That summer she spent mostly in California, learning from Fraenkel-Conrat and Robley Williams how to grow what she called "even more unnatural viruses," but also—to judge from the stream of postcards on which she communicated—having an extremely good time.

While she was in the West, I went to London, and occupied Rosalind's flat in Drayton Gardens. It is a small thing, though not a very small one, that she arranged by mail to have the kitchen stocked for me and that, when I arrived, I discovered with astonishment bordering on awe how very noticing Rosalind was, for everything in the refrigerator or on the shelves catered precisely to my likings: unsalted butter, Italian coffee, Hymettus honey, Devon cider, Gentlemen's Relish, Bath Oliver biscuits, Brie in a wedge—nothing missing, nothing wrongly chosen.* She came back in September; I returned from a visit

---

* But this sort of thing was not unusual with Rosalind. She had a thoughtfulness which I admired, but could never quite emulate. When a friend of hers came out of hospital with a newborn baby to find that her own house was uninhabitable, Rosalind took an impromptu holiday, installed mother, grandmother, and infant, in her flat, remembering while she was at it to provide a crib and a stack of diapers.

to Scotland to find her in the hospital, recovering from an operation which she dismissed as unimportant. Before I went back to New York, she and I went to a borrowed cottage for a few days; I fed her beefsteak to build up her constitution, took her on drives through the fen country, and then left her in London, fully convinced by her cheerfulness that whatever was wrong with her was minor, and cured.

It wasn't. We never met again. Rosalind wrote regularly until the end of 1957, mostly of joys and triumphs, never to complain. Shortly after I went back to New York she had a second operation; she went to stay with Francis and Odile Crick in Cambridge for her convalescence—they were good friends by then. She had traveled in Spain with them in the spring of 1956, plainly to everyone's pleasure. She wrote me to "please stop worrying . . . everything is going very well, and I expect to be fully back to normal some time next month." [1] She was apologetic about her illness: "It's really rather hard on relatives to put up with me twice in a month, but I've promised them I won't do it a third time." [2] I worried.

Later she stayed with Vittorio and Denise Luzzati, who were living at the time in Strasbourg, to recover from further surgery; but I never had further news of her health from Rosalind herself, only from other friends. She wrote of the pleasant honors which were beginning to come her way: a request from the Royal Society for an "exhibit of the structure of small viruses," another request to present her work at a Research Day at the Royal Institution,[3] a third to prepare a model of her TMV structure to figure as the central feature of the virus exhibition at the Brussels World Fair. She grasped at life while it was escaping her. During her 1956 visit to the United States, she had met a man whom she might have loved, might have married; she put this out of her mind, but she went on living, fiercely and even passionately.

In the last letter I had from her, she announced a forthcoming visit in the summer of 1958:

I am invited to a silly conference (Phytopathology!) at Bloomington, Indiana, in the last week of August.

This is a bad time for visiting people, especially as I should like to be at the biochemistry conference in Vienna in the first week of September. So I shall probably spend August and possibly part of July in the States, and thank you very much for the invitation, I should so love to stay with you in New York. . . . Another snag is that I have heard a rumour that this conference has accidentally invited more people than it has money for—and has offered expenses to all of them. However, I expect they'll sort that out.[4]

Whether the sponsors of the Phytopathology Conference sorted out their problems I don't know. D. L. D. Caspar represented her there at the meeting held four months after her death.

In her last months, Rosalind took up work on the polio virus, in the face of much opposition. Everyone around her kept pointing out that the stuff was deadly, and very highly infectious. She insisted upon being daring, perhaps in this instance out of bravado, for by then she was mortally ill, and perfectly aware that this was the case. She lived as normal a life as possible—dined out, entertained when she could, arranged a dinner party for her parents' fortieth wedding anniversary. Her last party was her father's birthday party.[5] And she worked tirelessly to the end.

There is no nice way of describing her dying, and there is not much use in trying. She knew from the autumn of 1956 onward that what she had was cancer; she told few people, asked for no sympathy. She fought every inch of the way. One of her close friends, Mair Livingstone, was a doctor—it was for her baby that Rosalind provided temporary housing, and diapers—and she provided her not only with the information that Rosalind demanded, but with the support she needed and could not ask for. She has written,

After her operation Rosalind told me that her surgeon had told her "how mad and bad and sad it was." No doubt Rosalind demanded to be told the full facts, but she was desperately hungry for comfort. . . .

She said that her surgeon, whom she had asked for a true prognosis, had been wholly discouraging. Rosalind had told the surgeon that she did not want to die, and the surgeon, who had strong religious faith, had exhorted her to appreciate that she was fortunate to know that she was dying and so to have the opportunity to prepare her soul for death. Rosalind was furiously angry about this interview. . . . I was ultimately able to make her laugh about it, and the "opaque mind of conscious virtue." . . . I encouraged Rosalind to accept future commitments which she thought good in themselves. The hospital at one point advised her not to travel out of reach lest she might need an emergency operation for intestinal obstruction. I told her to go ahead and risk the holiday she was longing for in Switzerland—at worst, I said, there were perfectly good, if expensive, surgeons in Zurich. By great good fortune nothing went wrong, and she enjoyed that holiday so much that she began to wonder whether there might indeed be a chance of getting better. By irregular degrees, however, she got worse. She could hardly bear the solicitude of her near family, and her remoteness must, I think, have been exceedingly hard for them. Towards the end she stayed with a brother and sister-in-law who succeeded in avoiding the tragic tone, and whose interesting family of children offered distraction. She never, to the end, lost her capacity to be amused, even when she was in pain it was possible to raise a smile by recounting some comic episode or observation. When she was in the Royal Marsden Hospital during the last days of her life, unable to eat anything and too feeble to lift her head, I told her an anecdote about my delicate and witty little boy Angus, then three years old. I expected a weak smile; to my astonishment she laughed aloud. I think that was the last time I saw her conscious.

I have known people in greater physical distress,

but never in greater anger about the unwanted, inconvenient, unjust and cruel sentence of dying young. She was indignant that there was not the technical skill available to avert death. She felt her mental power, and bitterly grudged its achievement being curtailed. She was saddened, but not depressed, I would say, since she remained combative to the end.[6]

Once Rosalind said to this friend, "Nobody else I can talk to about all this gives me any hope. You and I know that it is dangerous, but you don't advise me to stop everything because of that." [7] So she refused to stop anything; she worked and hoped and planned. She was coming to the United States in the summer of 1958; she had a visit to South America in mind. Muriel Franklin, her mother, says,

> It was just before her last fatal attack of illness that a letter came from the IVNIC Institute in Caracas, pleading with her to "honour them" by accepting for six months or a year or such time as she should decide a visiting fellowship at IVNIC University. Everything was to be provided for her, from living quarters to her travel expenses. It was a long letter, naïve, pressing and eager, obviously sincere, and it was wholly unexpected; it touched and amused her. Had she lived . . . she might have accepted just for the experience, for she loved travel and new ventures. For a time she kept the letter in her hospital room. But she knew only too well how slender was the chance that she could go. The letter was not answered in her lifetime.[8]

She fought death with stubborn courage, made plans for living when the plans were a mockery. She died as she had lived, with a passion for life that she never relinquished. On April 16, 1958, at the age of thirty-seven, Rosalind Franklin lost the battle.

To this day, I miss her.

Afterword

What more is there to say? Time passes; Rosalind has been dead for sixteen years; people die, and nothing stops. But it might be worth looking at a few of the things which have happened in those sixteen years. In 1962, the Nobel Prize for Medicine and Physiology was shared among Francis Crick, James Watson, and Maurice Wilkins, in recognition of their various and assorted contributions to work on DNA. Wilkins was then still at King's College, where he remains, directing its Medical Research Council unit; Watson moved on from Cambridge first to Cal Tech, then to a professorship at Harvard, and then to the biological laboratory at Cold Spring Harbor, which he heads. Crick stayed in Cambridge, and now is in charge of its MRC unit. All three have continued to work, Crick with great inventiveness and productivity; Watson produced an excellent textbook, *The Molecular Biology of the Gene,* before he turned his hand to *The Double Helix.*

But Rosalind is missing. How much she is missing is really very surprising. Certainly if she had still been alive when the Nobel Prize committee was considering its awards, she could scarcely have been overlooked, a claim I am willing to make on her behalf because J. D. Bernal, for one, made it long before I did.[1] Opinion is divided concerning how the prizes, in that case, would have gone: a two-way split between Crick and Watson? with Rosalind replacing Wilkins in the list of those honored? Nobel Prizes are not divided more than three ways, nor do they recognize the contributions of those unlucky enough to die before their turn comes. They do not matter as much as people think. Too many scientists with legitimate claims have been omitted in the course of the years, and no one with a passion for literature can be anything but skeptical about a prize denied to Leo Tolstoi but awarded to that good, kind woman Pearl Buck.

That Rosalind missed the Nobel list is no great cause for grief. But what troubles is the other lists she missed. Is it simply because sheer survival has pre-emptive claims that an encyclopedia gives her half a clause in an article on Bernal, simply to call her his pupil, which she was not, or in another half-clause in an article on Wilkins manages to do no more than associate her vaguely with a proudly recorded series of accomplishments [2]—having found the diameter and repeat distance of the DNA helix, having determined the density of DNA, having located the sugar-phosphate groups on the outside and the bases on the inside of the molecule—all of which are credited, bewilderingly, to Wilkins? Evidently these things were all worth doing, but to find out nowadays that it was Rosalind who did them takes some specialized effort. Is it because she failed to live to the age of forty-two that the DNA molecule exhibit in the natural history section of the British Museum omitted Rosalind from the list of people who had contributed to the discovery of the structure until complaints required a change? [3] If so, then to die young is crueler and more unjust than anyone supposed.

And this slow and gentle robbery does not stop. Linus Pauling, certainly a great scientist, and—one would imagine—a careful one, wrote an article for the DNA anniversary issue of *Nature* in which he, too, hands the credit for the B form photographs of DNA made by Rosalind over to Wilkins, and not once but twice.[4] The broad philosophical view—that it does not matter by whom a good thing was done as long as it was done at all—is a nice one which I fully appreciate; on the other hand, I would be slow to credit the a-helix to Smith, Jones, or Robinson when the literature of science tells me that it was Pauling who discovered it, and I have yet to find, anyway, that those who urge the broad, philosophical view with respect to the work of others are quite so broad or philosophical when it comes to the correct attribution of their own productions. Nor should they be. Those who ask—with lifted eyebrow— "Oh, do you work then for credit, for personal glory, and not for the sake of the work itself?" ask an idiot question. There is nothing disgraceful in claiming parenthood of your children, whether they are the children of your flesh and blood or of your labor and your intellect, and, indeed, if you look upon your sons and daughters as dispassionate contributions made for the sake of the continuance of the human race, then it is devoutly to be hoped that you are lying.

Those who do science work for the sake of science, but not as anonymous moles; those who write work for the sake of literature, but not impersonally either; and were either science or literature—or come to that, the fathering or bearing of children—to be stripped of the names of those who do their work, then responsibility vanishes with the credit. If one wishes to be philosophical, then let it be remembered that mankind can ill-afford that the works of man be irresponsibly done, and that credit and personal glory are no more in any case than the positive side of the story, on the negative side of which is punishment for failure and disgrace.

Rosalind has been robbed, little by little; it is a robbery against which I protest. And so this book has been written. Robert Frost said it better,

> Of all crimes the worst
> Is to steal the glory, . . .
> Even more accursed
> Than to rob the grave.[5]

Some of this robbery occurred when Watson published *The Double Helix*, though differently. Watson does not, in fact, deprive Rosalind of credit for her work; if one reads his book very carefully, and perhaps if one knows rather more than the average reader is likely to do about the dates and subjects of papers published in scientific journals, one can discover that he has a considerable respect throughout for who-did-what, and I praise him for it. It is very nearly the limit of my praise. It seems to me that he has carelessly robbed Rosalind of her personality, however, and this does not strike me as a virtuous act. I cannot read his motives, and I do not pretend to. Fictionalized, as he fictionalized her, Rosalind made a better character in a kind of highly personal, novelized memoir; but if one is going to create characters, I suggest that one is well-advised not to attach to them the names of real persons, living or dead.

He might well respond that what he wrote down about Rosalind was not fiction, but his view, Rosalind as he saw her. This would be a hard argument to answer—considering how widely, and wildly, opinions concerning any individual can vary—had he not thrown it away on his own behalf. In the epilogue to *The Double Helix,* in an addition which one reviewer described as "cloying," he took it all back. She was not, it appears, that way after all, but a good scientist, and throughout he was wrong. He—and Crick too—"came to appreciate greatly her personal honesty and generosity, realizing years too late the struggles that the intelligent woman faces to be accepted by a scientific world which often regards women as mere diversions from serious thinking." [6] What segment of the scientific world Watson inhabits is not made plain; surely there are others in which the custom of regarding women as *houris* meant to while away the leisure of the intellectual warrior is more unusual than usual. But never mind, he erred, and Rosalind, who is no longer "Rosy," was courageous, possessed of integrity, and admirable, as of course she was.

Confronted with this tribute, one is driven to respond ungratefully. "I see," one might say, "Rosy, the fictional character, was fiction. Rosalind was capable, intelligent, honest, and generous. But where did Rosy come from?" Watson's reply is predictable because he has made it often enough in conversation and in print: well, that was how he saw things back in 1952 and 1953, and he wrote *The Double Helix* strictly from the viewpoint of those years, with nothing in mind except the way things appeared to him then. This is a new literary form, an "as-if-it-were-true" memoir justified on the basis of "I-didn't-know-better-then." It is an interesting form; if available to all, innumerable monuments to past wrongheadedness may well be erected, with no correction, conversion, second thoughts, or better knowledge drawn from further experience required. All childish impetuosity of judgment may be unrepentantly repeated, and indeed, canonized. Erroneous first impressions are commonplace, incorrect first reactions are a universal human failing, but who before Watson has thought of elaborating them into a book intended to tell what really happened, what really was?

One has the right to ask why such a thing should be done at all. What were Watson's reasons? No one knows except Watson himself. He says he wants to tell how science is "done," and this is an admirable ambition. But how is it done? By using data unknowingly provided by others? That this occurred, that "Rosy" and everyone else at King's was unaware of what Watson and Crick possessed, he confesses to. In my experience of scientists, this is unusual behavior. I have taken the opinion of a great many working scientists on this question, and nobody yet has suggested that this is recommended behavior, that this is how science *should* be done.

Nobody denies, and I do not deny, that Watson and Crick found the structure of DNA, and a very beautiful piece of work it was. They, and no one else, deserve full credit for perceiving the nature of the base pairing; biologically speaking, this is what counts; and to have done this is in itself a very high and unarguable claim to glory. But as André Lwoff has put it, the evidence for the rest of the structure lay in Rosalind's data which they had received unorthodoxly and, therefore, surely they should either have confined their paper to the scheme of base pairing *or* offered her joint authorship for supplying the rest of the information.[7] To have failed to do this was, then, to claim more than was original with them— how science is done, perhaps, but not with the warmest approval of the community in which science exists.

It is not really a minor thing. If one is trying to puzzle out not only why Watson wrote *The Double Helix*, but why he wrote it as he did, it is quite a major consideration. Obviously it is embarrassing to put in a telephone call from Cambridge to London in order to explain that one has unorthodoxly obtained, and made use of, data which originated at King's; it is more embarrassing still to do this when one is connected with an organization which has an unabrogated agreement with King's College to refrain from working on DNA. It is most embarrassing of all to think of calling one who has behaved in a friendly manner, who has offered confidences—Wilkins—to say what the friendliness has led to, what the confidences have produced. The embarrassment is understandable, and no one could be so heartless as to fail to sympathize with it. But the gentlest heart will harden somewhat when it realizes that alternatives existed which would have minimized the embarrassment while still meeting fully the ethical requirements which make intellectual life possible. A wholly original paper on the base-pairing scheme might have been written and published, wholly by Crick and Watson, and a brilliant and insightful paper it would have been, too, quite enough of both to insure them lasting fame. A joint paper embodying the whole structure might have been written in which the contributions of Crick and Watson and of Rosalind Franklin would have been accurately and wholly defined, and though the glory attached to discovering the structure would then have been somewhat more divided, history would never have been confused concerning precisely who-did-what. There was, indeed, glory enough to go around.

Neither of these alternatives was adopted. To be honest, the glory was hogged. Well, so it happens sometimes, but when it does, who defends it? Is there a defense in *The Double Helix?* Certainly much of Watson's book is a nice picture of the author as a lovable rogue, but some of it is a nice argument that suggests that where you lack love for a bright, uppish, contentious female whose hair-do does not appeal, you can scarcely be blamed for taking whatever advantage of her that is available. Nobody loves a bluestocking, nobody loves an unfeminine female who quarrels with her boss, who flatly contradicts the men around her, who insists upon the rightness of her own notions. Therefore, Rosalind becomes "Rosy," an unfeminine bluestocking meant to "assist" Wilkins and not to argue with him, to be meek and helpful and not argumentative.

She was in what we may call "real life" none of these things. Never mind. The average reader—a target at whom Watson aimed *The Double Helix* with very great accuracy—will never know the difference. The average reader will buy the package, and come up in the end full of sympathy. The ethics of science, then, become roughly the same as those of used-car dealers. If you have a secretary around the office who is like that Rosy woman you are perfectly justified either in exploiting her or firing her. Was this why "Rosy" was invented? To rationalize, justify, excuse, and even to "sell" that which was done that ought not really to have been done?

It is a guess, no more; it is not a total improbability. It is interesting, and somewhat suggestive, to realize that "Rosy" in all her unacceptability was intended to stand uncorrected; Watson admits freely that the tribute in the epilogue was pressed upon him.[8] It is also interesting, and no less suggestive, to realize that, if Rosalind had lived until 1968, *The Double Helix* could scarcely have been published in the form in which it was. Rosalind was not a meek creature, and she was utterly honest—Aaron Klug, who knew her far better than Watson ever did, speaks of her as the most honest person he has ever known, and my own description of her would be identical to his. Had she read *The Double Helix*, and realized what she never did realize in her lifetime, that her work was appropriated and used without proper credit, I doubt very much if she would have laughed, and considered it a clever joke on the part of "honest Jim." [9] Nor is it likely that anyone, however insensitive, would quite dare to create such a picture of a living person as the one Watson created in his character of "Rosy," for living people have means of defending themselves which are denied to the dead.

None of this entirely escaped all reviewers of *The Double Helix*, though much of it escaped some of them. Lwoff's review in *Scientific American* is, for instance, uneasy about the appropriation of Rosalind's data: "It is a highly indirect gift which might rather be considered a breach of faith." He also remarks that "His portrait of Rosalind Franklin is cruel. His remarks concerning the way she dresses and her lack of charm are quite unacceptable. At the very least the fact that all the work of Crick and Watson starts with Rosalind Franklin's X-ray pictures and that Jim has exploited Rosalind's results should have inclined him to indulgence." [10] All that I suggest is that perhaps the lack of indulgence and the cruelty were not unconnected with the exploitation. It is after all so very much easier to indulge happily in breaches of faith with those we can convince ourselves we despise.[11]

If all this ended with Rosalind, if *The Double Helix* and, in particular, the events which it gleefully describes, had no consequences beyond an injustice to one individual now dead, that would be sad. What Rosalind has been deprived of she is unaware of; and, though it has always been thought bad taste, to say the least, to dance upon graves, this too has happened before. What is more than sad, and worse than bad taste, is that there is no telling how far the precedent that has been set may be carried. Science is a competitive business, and when the rules which for years have worked fairly well to keep the competition civilized begin to crumble, and are applauded by success for crumbling, competition may become unrestrained. The rush to publish before someone else publishes —the "Linus would get there first" attitude—is a rush in which poor work done in haste is encouraged. Not everyone hurrying to the golden goal is a Crick or a Watson, after all, but vastly inferior young intellects may just the same take them as inspiration and see no reason why whatever is portable ought not to be picked up along the way, regardless of to whom it belongs.

A generation of graduate students in science read *The Double Helix* and learned a lesson: the old morality was dead, and they had just been told about its demise by a respected, highly successful Nobel Laureate, an up-to-date hero who clearly knew more about how science was acceptably "done" than the old-fashioned types who prattled about ethics. One of them told me cheerfully that the way to get on was to keep your mouth and your desk drawers locked, your eyes and ears open, and "then beat the other guy to the gun." No doubt there have always been ambitious graduate students—and

postgraduates, too—who thought this way; few of them announced it; none of them thought that such engaging frankness
would be a recommendation. They have learned differently.
Another graduate student said that it was all down in *The
Double Helix,* how to get ahead, and nobody thought the
worse of Watson, did they? [12]

Jobs for young scientists are not at present as plentiful by
any means as they were ten years ago, and in the meantime a
very great number of young scientists have been produced by
the universities; competition is severe nowadays, and getting
ahead is bound to be difficult. If it depends upon one's own
output and integrity, it may be very difficult, indeed. But why
let it depend upon these things? A method of succeeding very
rapidly and effectively has become an advertised product. Read
all about it in confessional, unrepentant print. I have lived a
good many years in the world of scientists, and it is a world I
have grown to love; I have learned in the course of doing research on this book that there are elements in it, and growing
ones, which I am learning to hate, and with bitterness.

There seems to me almost no way in which a good and
greatly gifted person—Rosalind Franklin, now dead—can be
used for wrong and embittering purposes that she has not been
put to, helplessly and virtually undefended. Certainly she has
been used, thanks to *The Double Helix,* to menace bright
and intellectually ambitious girls. I went once to a public
meeting of a local school board and heard a man stand up to
demand that science requirements for girls be dropped from
the high school curriculum because he had a daughter, and
he "didn't want her to grow up like that woman Rosy-what's-
her-name in that book." I think I wept. It was not much con-
solation to know that the high-school curriculum is fixed beyond the meddling of local boards at the demand of local
extremists. But that man has a daughter, for all I know an intelligent and gifted one, and I do not really like to contemplate
her future.

One neurotic man overstimulated by his reading may not
make a trend, but if the trend fails to establish itself, it is not
because Watson has put anything in its way. Certainly Rosa-

lind has been used—vastly warped to fit the purpose—to provide reasons why men who work with intelligent women should resent them. No opportunity that I can find was missed in *The Double Helix* to emphasize this point. Certainly she has been taken up, in the Watson version, by antifeminists everywhere—presumably Watson is one of them; if he isn't, he has scarcely made this clear—until women everywhere who think (it is not an extreme attitude) that work opportunities and rewards should be connected to ability rather than to sex find themselves apologizing for her.

She was no one who ever needed to be apologized for by anyone. She was over and over again a victim of the sort of thinking that not only prefers women to confine themselves to kitchen and nursery and possibly church, but is outraged by their presence anywhere else at all; she suffered from this often and long, in its subtle forms as well as its overt ones; she bore with it not always calmly, not always meekly, not always sweetly, but always with dignity. It is nothing to be calm or meek or sweet about; and it is endlessly to her credit that she let no such opposition, no such obstacles, get in her way. She was a very good scientist and a very productive one, a very honest one of unimpeachable integrity, and she was not the less of any of these things because she was a woman, and often opposed on no better grounds than her sex.

This is information, now lost, which ought to be restored to public awareness. The notion that she was an ineffective scientist has become widespread; it is much reinforced by the mysterious way in which the work she did has a tendency to be absorbed by other people in recent historical or encyclopedic coverage. It has much to do with *The Double Helix*, too. That is a book which can be read by everyone—the only other praise I have available for Watson is for his readability —and for a time it appeared to be read by everyone; even now there are high schools which use it as a collateral text.

It leaves a certain distinct impression of Rosalind's abilities which can only be fully corrected by people who know enough about science to go to the places where the original sources exist, and read them, and understand them. There are not many such people. Even so intelligent, perceptive, and sympathetic a writer on the topic as Elizabeth Janeway was essentially required to take the facts on Watson's word and, therefore, to write about "Rosalind Franklin, a capable (if sometimes mistaken) research scientist in the King's College (London) team headed by Maurice Wilkins." [13] Rosalind was much more than capable; in the matters of which Mrs. Janeway writes she was not mistaken but quite correct; and the team headed by Wilkins she was not part of; but *The Double Helix* puts it differently, and to find out the facts requires—it has, anyway, required me—something on the order of three years of research.*

I think that some, at least, of the persistent undervaluation of Rosalind's powers and contributions may be connected with the rise of modern molecular biology, a new science which does not please all scientists quite as much as it pleases *TIME*,[14] or the popular press. Molecular biology has grown very much upon model building techniques; Erwin Chargaff—another person whose contributions to the discovery of the structure of DNA have not precisely been quite fully appreciated—writes about this uneasily,

> I remember vividly my first impression when I saw the two notes on DNA that appeared in *Nature* 21 years ago.† The tone was certainly unusual: somehow oracular and imperious, almost decalogous. Difficulties, such as the even now not well-understood manner of unwinding the huge bihelical structures under the conditions of the living cell, were brushed aside, in a Mr. Fix-It spirit that was later to become so evident in our scientific literature. It was the same spirit that soon brought us the "Central Dogma" to

---

* They would, of course, have taken Watson much less time to establish. His book, however, does not deal in facts, but in his impression of the facts.

† He cites: Watson, J. D., and Crick, F. H. C., *Nature* 171 (1953): 737, and *Nature* 171 (1953): 964.

which I believe I was the first to register my objection,
never having been very fond of gurus with a Ph.D.
I could see that this was the dawn of something new:
a sort of normative biology that commanded nature
to behave in accordance with the models.[15]

Whether Chargaff's uneasiness is or is not justified I am in
no position to decide. What I do know is that the model
builders are insistent, that *The Double Helix* is in part—it is
evidently a much more complex book than it appears on the
surface to be—a hymn in praise of model building techniques,
and that, while it is natural to love one's own methods and the
successes they have achieved for one, the hymning approach
produces somewhat exaggerated results. Crystallographers who
work with diffraction patterns and mathematical analysis are
rigid and unimaginative, according to Watson. Whether they
are or are not is debatable, considering the successes they have
achieved, and considering that something over 90 percent of
the structures worked out to date have resulted from their non-
model-building methods.

Rosalind was re-created by Watson into one of these rigid
creatures, a highly trained crystallographer equipped by long
education with a narrow and cumbersome approach. She was
so little a crystallographer in any traditional or well-defined
manner that she never called herself one. But never mind. She
didn't build models. Embarrassingly, she provided—however
unwittingly—out of her non-model-building experimental approach, at least half the essential information out of which the
successful model was built; but she didn't build models. If one
is truly dedicated to the proposition that in the hands of model
builders lie all the solutions to all biological puzzles, it is more
than embarrassing to acknowledge how dependent the model
was—and is—upon other methods entirely. It is, indeed, close
to heresy. Heretics, as we all know, should be burned; lacking
a handy auto-da-fé, they can at least be damned as ineffective.

So Rosalind, who was in science remarkably pragmatic, remarkably open to using whatever methods or approaches
looked to her like the most useful in prying open the shell of

the problem, remarkably flexible in her techniques, and remarkably successful in the techniques she used, is transformed
into the rigid opponent of model-oriented molecular biology—
not a true believer and, therefore, an ineffectual, mistaken
scientist. This element of *The Double Helix*, as propaganda for
a method, is of course scarcely obvious to the reader who
neither knows nor cares whether models are built or are not
built; it was scarcely obvious to me until the monotonous cry,
She didn't build models, began to appear as a rather noisy
way of burying what she did do.

What she did do was fine science. What she was, as a person,
was both noble and lovable. My pride is that she was my
friend.

Notes

ONE *An Introduction*

[1] James D. Watson, *The Double Helix* (New York: Atheneum, 1968).
[2] Ibid., p. xii.
[3] Ibid., p. xii.
[4] Ibid., p. 17.
[5] Ibid., p. 17.
[6] Elizabeth Janeway, *Man's World, Woman's Place: A Study in Social Mythology* (New York: William Morrow, 1971), p. 102.
[7] Watson, *The Double Helix*, p. xii.
[8] New York *Times*, April 17, 1958.

TWO *Rosalind*

[1] Muriel Franklin, *Rosalind* (privately printed), pp. 5–6.
[2] Stanley Jackson in the *Jewish Chronicle* (London), November 27, 1970.
[3] Helen Bentwich, *If I Forget Thee: Some Chapters of Autobiography, 1912–1920* (London: Elek, 1973), p. 2.
[4] Ibid.
[5] Ibid., p. 3.
[6] Muriel Franklin, *Portrait of Ellis* (privately printed), p. 156.
[7] Ibid., pp. 167–168.
[8] Virginia Woolf, *Three Guineas* (London: Hogarth Press, 1938), p. 55.
[9] Ibid., p. 277.
[10] A brief history of women's education in Britain may offer a little, though not much, illumination. The two women's colleges at Cambridge, Girton and Newnham, were founded in 1869 and 1873 respectively; between 1879 and 1893, five women's institutions were

established at Oxford. Oxford granted women degrees—real ones, not "titular"—in 1921; this privilege was not granted at Cambridge until 1947.

For comparison, one might observe that London University accepted women without reservations in 1878, apart from study in medical schools. In the United States, the first degrees for women were awarded by Oberlin in 1841. Before 1860 at least two American state universities accepted women on equal terms with men. Vassar was empowered to grant degrees in 1861, Wellesley in 1870, and Bryn Mawr in 1885. Women joined men at Cornell in 1865; Radcliffe was "annexed" to Harvard in 1879, but its graduates had to wait until 1943 for the special blessing of a Harvard degree.

The laggard's pace of Oxford and Cambridge in following these examples is baffling. It has often been suggested that the cause of women's education was, at these two places, unfortunately mixed in with, and consequently held back by, another cause, that of education for men of the middle and lower classes. The issue was that of entrance into Oxford and Cambridge on the basis of competitive examinations, rather than by the methods of selection which had hitherto prevailed. A resistance to the entrance of women by examination might, therefore, serve as an argument against entrance by examination by anyone, and indeed, as a dislike of brainy women is a more popular social prejudice than an admitted dislike of brainy but otherwise deprived young men, such an argument might well have been very useful. It is true that as both Oxford and Cambridge have gradually become more and more open to applicants on the basis of competitive examination, until at present there is virtually no other means of admission, they have also become gradually less and less hostile to the presence of women. Cambridge has recently added two new women's colleges to its foundations, and at both the senior universities, a number of colleges formerly restricted to men are now coeducational.

[11] Dorothy L. Sayers, *Gaudy Night* (London: Victor Gollancz, 1935), p. 175.

[12] Ibid., p. 260.

[13] Letter from REF to parents, March 16, 1938.

[14] Letter from REF to parents, October 12, 1940.

[15] Ibid.

[16] Letter from REF to parents, undated, 1940.

[17] Letter from REF to parents, undated, 1940(?).

[18] Letter from REF to parents, October 24, 1940.

[19] Rosalind's claustrophobia was not acute, but it was real. She concealed it habitually. She mastered the bus systems of several cities in order to stay out of subways and, in the later stages of the war, became an air raid warden, partly because it gave her an excuse to stay out of shelters. She shared this duty with her cousin Irene Franklin (now Mrs. Neuner) who has written, "One night a stick of bombs fell. We were neither on duty, but thought we had better go and help. . . . We could see a direct hit on a house . . . also the lights and gas had gone out. . . . We passed another direct hit and stopped to investigate. Planes were still around, and there was a lot of shrapnel coming down. Ros went to the post to report, and I went into the undestroyed half of a . . . house and got the people out of the cellar where they were hiding. Swapping notes afterwards, Ros said that she was so grateful that I had gone into the house, and she was too scared—and I was glad that she had walked across the open Common."

[20] Communication from R. G. W. Norrish, September 22, 1970.

[21] Rosalind was often associated with, but never uncritical of, left-wing political thinkers. Her college-day opinions are indicated by two letters she wrote home to her parents from Cambridge. In one she described an acquaintance as full of "various and numerous *Daily Worker* facts and theories" (November 30, 1939); in December 1940, she wrote, "The Communists here . . . defend Russia vigorously. They make more excuses, reason and provocations than Russia herself ever dreamt of. I should not have seriously believed anyone would talk as they do if I had not heard them."

The skepticism so expressed never materially changed. In 1952, during a trip to Yugoslavia, she wrote to this author praising the scenery, the scientists whom she was meeting, and many other things, but she added that she had also heard "the most preposterous Communist-propaganda statements."

Her politics looked leftish to the conservative, and nowhere nearly left enough to those who could be called radical. The fact is that Rosalind's interest in politics was mild, on the level of that of the ordinary citizen. After she returned to England from France in 1951, she canvassed for Labour candidates in two elections, which is her only recorded adult personal political action; actively, then, she supported moderate parliamentary socialism. In a living-room debate, she might often have gone further one way or the other, depending upon how much fun there was in the argument.

[22] BCURA report, 1942.

[23] These papers are:

Bangham, D. H., and Franklin, R. E., "Thermal expansion of coals and carbonized coals," *Trans. Farad. Soc.* 42 B (1946): 289–294.

Franklin, R. E., "A note on the true density, chemical composition, and structure of coals and carbonized coals," *Fuel* 27, no. 2 (1948): 46–49.

Franklin, R. E., "Study of the fine structure of carbonaceous solids by measurements of true and apparent densities. I. Coals," *Trans. Farad. Soc.* 45 (1949): 668–682.

Franklin, R. E., "Fine structure of carbonaceous solids by measurements of true and apparent densities. II. Carbonized coals," *Trans. Farad. Soc.* 45 (1949): 668–682.

Bangham, D. H.; Franklin, R. E.; Hirst, W.; and Maggs, F. A. P., "A structural model for coal substance," *Fuel* 28 (1949): 231.

The title of Rosalind's thesis was, "The Physical Chemistry of Solid Organic Colloids with Special Relation to Coal and Related Materials."

[24] Communication from Professor Peter Hirsch, Isaac Wolfson Professor of Metallurgy, Oxford.

THREE *Paris*

[1] Gallant, Mavis, "Annals of Justice," *New Yorker*, June 26, 1971.

[2] Ibid.

[3] The papers published by Rosalind Franklin on carbons other than coal are:

Franklin, R. E. (Lab. Centr. Serv. Chim. Etat, Paris), "Influence of bonding electrons on the scattering of X-rays by carbon," *Nature* 165 (1950): 71–72.

Franklin, R. E. (Lab. Centr. Serv. Chim. Etat, Paris), "The interpretation of diffuse X-ray diagrams of carbon," *Acta Cryst.* 3 (1950): 107–121.

Franklin, R. E. (Lab. Centr. Serv. Chim. Etat, Paris), "A rapid approximate method for correcting low-angle scattering measurements for the influence of finite height of the X-ray beam," *Acta Cryst.* 3 (1950): 158–159.

Franklin, R. E., "The structure of carbon," *J. chim. phys.* 47 (1950): 573–575.

Franklin, R. E., "Graphitizable and nongraphitizable carbons," *Compt. rend.* 232 (1951): 232–234.

Franklin, R. E. (Lab. Centr. Serv. Chim. Etat, Paris), "Crystallite growth in graphitizing and nongraphitizing carbons," *Proc. Roy. Soc.* A 209 (1951): 196–218.

Franklin, R. E. (Lab. Centr. Serv. Chim. Etat, Paris), "The structure of graphitic carbons," *Acta Cryst.* 4 (1951): 253–261.

Franklin, R. E. (Birkbeck College, London), "Graphitizing and nongraphitizing carbon compounds. Formation, structure and characteristics," *Brennsdorf-chem.* 34 (1953): 359–361.

Franklin, R. E., "The role of water in the structure of acid," *J. chim. phys.* 50 C (1953): 26.

Franklin, R. E. (Birkbeck College, London), "Homogeneous and heterogeneous graphitization of carbon," *Nature* 177 (1956): 239.

Franklin, R. E., and Bacon, G. E., "The alpha dimension in graphite," *Acta Cryst.* 4 (1951): 561–562.

Watt, J. D., and Franklin, R. E., "Changes in the structure of carbon during oxidation," *Nature* 180 (1957): 1190.

A summary of Rosalind Franklin's work in carbons may be desirable. Her special talent lay in her quick recognition of similarities between apparently unlike materials. She was the first to perceive that coals and some organic chemicals, among which she later included plastics, often have characteristics in common. She was able to divide coals, plastics, and some other solid organic materials into two principal categories: those which, upon heating, yield nongraphitizing carbons (low-rank coals, substances rich in oxygen or poor in hydrogen, polyvinylidine chloride; these carbons have low density, large fine-structure porosity, and are very hard) and those which convert readily to graphite (coking coals, substances rich in hydrogen, polyvinyl chloride; these form carbons which are soft, compact, and of high density).

Within a few years, the discovery of the nongraphitizing carbons was to have important industrial consequences. About 1950, workers at the Bell Telephone Laboratories became interested in the polymer carbons of the type Rosalind had pioneered. Industrial development of solid shapes that could be made from polymer carbons was taken up around 1960 in Japan, England, and probably elsewhere. The products are known variously as glassy or vitreous carbons, and are manufactured in the form of crucibles, tubing, and other articles for specialty application.

The unique behavior of the nongraphitizing carbons could be attributed to the formation of a strong system of cross-linking between the carbon crystallites which, together with the fine-pored nature of the structure, prevented rearrangement of the material into graphite. At high temperatures, the pores of nongraphitizing carbons shrink to such minute size that even helium atoms are unable to enter them.

In a series of brilliant papers which emerged from the Paris association, Rosalind dealt with the interpretation of the X-ray studies of the nongraphitizing carbons as well as of those which transform readily to graphite. At the same time, she made notable advances in X-ray diffraction methods for determining the structures of large, complex substances, and developed the accompanying mathematical techniques. (The information upon which this analysis is based has been kindly provided by Dr. Einar Flint.)

To this should be addded the following comment made by J. D. Bernal in the London *Times* shortly after Rosalind's death:

> She discovered in a series of beautifully executed researches the fundamental distinction between carbons that turned on heating into graphite and those that did not. Further she related this difference to the chemical constitution of the molecules from which carbon was made. She was already a recognized authority in industrial physico-chemistry when she chose to abandon this work in favour of the far more difficult and more exciting fields of biophysics (April 19, 1958).

FOUR *The Problem*

[1] Friedrich Knipping and von Laue, *Bayerische Akademie den Wissenschaften* (1912), pp. 303–322.

[2] "The diffraction of short electromagnetic waves by a crystal." See W. L. Bragg, M.A., F.R.S., Cavendish Professor in the University of Leeds, Bakerian Lecture, reported in *Phil. Trans. Roy. Soc.* A 215 (1915).

[3] See Chapter 3, note 3.

[4] The reader is here recommended to seek further information on the development of molecular biology in Gunther S. Stent, *Molecular Genetics: An Introductory Narrative* (San Francisco: W. H. Freeman, 1971).

[5] Ibid., pp. 184–185.

FIVE *"One Cannot Explain These Clashes of Personality"*

[1] A remark made to the author by Sir John Randall in an interview.

[2] Quoted from an interview with Raymond Gosling.

[3] Quoted from an interview with Sir John Randall.

[4] Quoted from an interview with Raymond Gosling.

[5] Quoted from an interview with Maurice Wilkins.

[6] Quoted from an interview with Raymond Gosling.

[7] Quoted from an interview with Raymond Gosling.

[8] Quoted from an interview with Raymond Gosling.

[9] Quoted from an interview with Maurice Wilkins.

SIX *The Making of a Discovery*

[1] See Peter Pauling, "DNA—the race that never was?" *New Scientist,* May 31, 1973.

[2] I do not invent this necessity to give proper credit for ideas developed conversationally. In *The Double Helix,* Watson faithfully reports on pages 57 and 58 the degree to which Crick was upset when Bragg apparently used, in a paper, an argument propounded by Crick some months earlier, and not appropriately acknowledged. What Crick wanted was not an explanation but an apology. Sir Lawrence Bragg denied any knowledge of Crick's notions, and for his part, was furious at the implication that he had used the ideas of another scientist in an unforthright manner. But Crick was not persuaded. It does not matter for present purposes whether in this instance Bragg was right or Crick was right. What is much to the point is that Crick was reasonably sensitive on his own behalf to the need to defend his originality, and Watson was his strong supporter.

[3] Watson, *The Double Helix,* pp. 15–16.

[4] Ibid., p. 16.

[5] Ibid., p. 98.

[6] The relevant passage in *The Double Helix* is found on p. 172 ff. It raises again the specter of Linus Pauling as the chief competitor in the DNA race, and deals with Bragg's agreement to Watson's and Crick's plans to build a model in opposition to Pauling. This is, however, somewhat misleading. Pauling had proposed a structure for DNA which was obviously incorrect; there was no reason to think that he was about to produce another, or even that he was pursuing the problem very vigorously. In *The Double Helix,* Watson frequently refers to information passed on from Linus Pauling

to his son Peter Pauling, and then to Watson, which led Watson to believe the contrary. This Peter Pauling disputes in his article, "DNA—the race that never was?" where he denied having engaged in this friendly espionage at all. His comment on the "race" is,

> The only person who could conceivably have been racing was Jim Watson. Maurice Wilkins had never raced anyone anywhere. Francis Crick liked to pitch his brains against difficult problems. To my father, nucleic acids were interesting chemicals, just as sodium chloride is an interesting chemical, and both presented interesting structural problems. For Jim, however, as a geneticist, the gene was the only thing in life worth thinking about and DNA was the only real problem worth tackling.

The point is worth making, because Watson refers insistently to Pauling as the person to be beaten in a neck-to-neck race, and does not refer to any real competition from King's. But Pauling is an immensely successful man, twice a Nobel Prize winner, a scientist of gigantic reputation. There is something pleasing, and touching, in the thought of two young men jauntily defeating this giant. The person who was very close to producing the correct structure was Rosalind Franklin—a fact never quite conceded in *The Double Helix*, but nevertheless demonstrably true. The thought of two young men jauntily defeating this young woman is not so touching or pleasing.

One has earned the moral right to comment of this nature because *The Double Helix* so largely consists of personal, psychological commentary which neither appeals to, nor is supported by, the body of objective evidence. It will be seen that Watson drew onto his side of the game, if that is what it was, as many allies as possible, whether they were voluntary or involuntary. Certainly he appears to have misrepresented the circumstances in which Max Perutz provided information from King's via the Medical Research Council biophysics committee of which Perutz was a member; and against this Perutz defends himself ably in a letter published in *Science* 164 (June 27, 1969): 1537–1538.

Equally, the existence of internal squabbling at King's is given by Watson as a reason why Bragg, as director of the Cavendish, withdrew his objections to Crick's and Watson's work on DNA. A hint is also incorporated to the effect that it was recognized in professional circles that no solution to the DNA problem could possibly have emerged from King's because of internal divisiveness. There is simply no proof to show that Bragg acted on such grounds, or that such grounds existed. At no point does Watson's representation of the situation hold water.

SEVEN *"She Was Definitely Antihelical"*

[1] Quoted from an interview with M. H. F. Wilkins, June 15, 1970.
[2] From *Interim Annual Report: Januaray 1, 1951–January 1, 1952,* written by Rosalind Franklin, dated from the Wheatstone Physics Laboratory, King's College, London, February 7, 1952.
[3] Ibid.
[4] From an interview with Raymond Gosling, May 18, 1970.
[5] From an interview with M. H. F. Wilkins, June 15, 1970.
[6] W. T. Astbury was among the earliest workers on DNA. He obtained the first X-ray diffraction pictures of the substance.
[7] From *Interim Annual Report.*
[8] Watson, *The Double Helix,* pp. 69–72.
[9] Ibid., pp. 69–70.
[10] From an interview with M. H. F. Wilkins, June 15, 1970.
[11] Watson, *The Double Helix,* p. 69.
[12] Ibid., pp. 69–70.
[13] Ibid., p. 94.

EIGHT *"On the One Hand a Defeat,*
On the Other a Triumph"

[1] A comment made on Rosalind's DNA work by Aaron Klug.
[2] A letter from Rosalind Franklin to the author, dated March 1, 1952.
[3] Watson, *The Double Helix,* pp. 17–18.
[4] An opinion expressed to the author by Sir John Randall in an interview, June 15, 1970.
[5] An opinion expressed to the author by M. H. F. Wilkins in an interview, June 15, 1970.
[6] Interview with Randall, cited above.
[7] A letter from Rosalind Franklin to the author, written on "the boat from Split to Rijeka," dated June 2, 1952.
[8] Rosalind enjoyed herself thoroughly in Yugoslavia. She made several friends there; one of them, Dr. Katarina Kranjc, has provided the following sketch:

I was working on my thesis on small-angle scattering, and I applied to Rosalind for some reprints of her carbon work. Perhaps she remembered my name, and so the organizers asked me to join them in meeting Rosalind at the railway station when she was arriving [in Zagreb] from Ljublana. I was happy about this duty because I was only a beginner in scientific work, and knew that Rosalind had already obtained fundamentally new results on carbons. . . . Thus I took a large bunch of red roses to the station, and I had the impression that she was pleased with them.

I do not know what I imagined a woman scientist should look like, but I certainly did not expect her to be such a charming young girl: my astonishment was enormous. Later I realized that she always looked much younger than her actual age" (letter from Dr. Katarina Kranjc to the author, March 24, 1970).

[9] Watson, *The Double Helix,* p. 167.

[10] Rosalind saw Vittorio Luzzati on several occasions, and they corresponded frequently. He appears to have been the only scientist who followed what she was doing, and who talked to her about it at length. It has been suggested that Rosalind's concentration for a time on the A form, and her studious application to it of Patterson superposition methods—which in this case were not very fruitful—were a result of Vittorio's influence. This can at best be a half-truth. Rosalind was never easily influenced where unconvinced. But it is testimony to her isolated position. The disadvantages of being essentially a lone worker can hardly be overestimated.

[11] Watson, *The Double Helix,* p. 162.

[12] Ibid., p. 165.

[13] Ibid., p. 99.

[14] Ibid., p. 165.

[15] Rosalind E. Franklin, and R. G. Gosling, *Nature* 172 (July 25, 1953): 156.

[16] Wilkins expressed himself clearly about this during an interview on June 15, 1970.

It was all here. They were working at Cambridge along certain lines, and we were working along certain lines. It was a question of time. They could not have gone on to their model, their correct model, without the data developed here. They had that—I blame myself, I was naïve—

and they moved ahead. Put it this way, if they were out of the picture entirely, we would still have got it, though it would have taken a bit longer. If we were out of the picture, if they hadn't got our stuff, they'd have had to develop it, and that would have taken time—I don't know how long, I think longer still. We were scooped, I don't think quite fairly.

It is hard to argue with Wilkins here. His points are very well taken. He cannot be blamed if he is indignant that what he had to say in normal conversation with another scientist should have turned out to be an indiscretion. It is difficult to say exactly how long Crick and Watson would have needed to produce data of equivalent value to those they obtained from King's, but it is very probable, indeed, that Wilkins guesses correctly as to relative time.

[17] The date has been suggested by Robert Olby. See his "Francis Crick, DNA, and the central dogma," *Daedalus* 99, no. 4 (Fall 1970), *Proc. Am. Acad. of Arts and Sciences:* 960–961.

[18] Watson, *The Double Helix*, pp. 167–169.

[19] Interview with Maurice Wilkins, June 15, 1970.

[20] See Rosalind Franklin, *Interim Annual Report, January 1, 1951– January 1, 1952.*

[21] Max Perutz, communication published in *Science* 164 (June 27, 1969): 1537–1538.

[22] Watson, *The Double Helix*, pp. 181–182.

[26] Max Perutz, communication in *Science*.

[24] Ibid.

[25] J. D. Watson, communication in *Science* 164 (June 27, 1969): 1539.

[26] Max Perutz, communication in *Science.*

[27] Robert Olby, "Francis Crick."

[28] Watson, *The Double Helix*, p. 169.

[29] Ibid., pp. 176–178.

NINE *Winner Take All*

[1] J. D. Watson and F. H. C. Crick, "A structure for deoxyribose nucleic acid," *Nature*, no. 4356 (April 25, 1953), p. 737.

[2] Ibid.

[3] M. H. F. Wilkins, A. R. Stokes, and H. R. Wilson, "Molecular structure of deoxyribose nucleic acids," *Nature*, no. 4356 (April 25, 1953), p. 738.

[4] Rosalind E. Franklin and R. G. Gosling, "Molecular configuration in sodium thymonucleate," *Nature,* no. 4356 (April 25, 1953), p. 740.

[5] Watson, *The Double Helix,* pp. 125–126.

[6] Robert Olby, "Francis Crick, DNA, and the central dogma," *Daedalus* 99, no. 4 (Fall 1970), *Proc. Am. Acad. of Arts and Sciences:* 958.

[7] Watson, *The Double Helix,* p. 192.

[8] Ibid., p. 160.

[9] Robert Olby, "Francis Crick," p. 960–961.

[10] Watson and Crick, "A structure for deoxyribose nucleic acid," p. 738.

[11] Ibid., p. 737.

[12] Watson, *The Double Helix,* p. 201.

[13] From a letter from Robert Olby to Aaron Klug.

[14] Draft paper dated 17/3/53, with corrections, additions, and mathematical calculations added in Rosalind's hand. (In the possession of Dr. Aaron Klug).

[15] Aaron Klug, "Rosalind Franklin and the discovery of the structure of DNA," *Nature* 219 (August 24, 1968): 808 ff.

[16] Watson, *The Double Helix,* p. 210.

[17] Ibid.

[18] Ibid., p. 212.

[19] Ibid.

[20] For an example of the first, see L. D. Hamilton, "DNA: Models and reality," *Nature* 218 (May 18, 1968): 633 ff.

For an example of the second, see *TIME,* April 19, 1971, pp. 34 ff.

[21] A note is demanded here. Francis Crick has been very generous with his time and his opinions, a generosity for which I am deeply grateful. I refrain from reporting all his opinions about individuals, but I think his opinion that he regards *The Double Helix* as a "contemptible pack of damned nonsense" ought not to be suppressed. He has expressed the same opinion to others in similar terms.

The following quotations are taken from the author's interview with Crick in Cambridge on June 16, 1970.

Concerning the matter of why he and Watson undertook to work on DNA when it was by arrangement a project of the King's College laboratory, Crick said, "They were mucking it up at King's, getting nowhere. Don't you see, if I hadn't done something about it, Pauling would have got it out first. I know Linus was wrong in his first guess, but Linus isn't stupid. . . . He'd have done it."

When asked whether, then, he believed that no one at King's would ever have solved the problem, Crick said, "Oh, don't be silly. Of course, Rosalind would have solved it. . . . With Rosalind it was only a matter of time."

I do not try to reconcile these statements.

In Crick's expressed view, the problem was susceptible to X-ray diffraction methods "if anyone knew how to use them, which Rosalind did. But it's slower than model building, and she wouldn't build models. . . . It was all there. [Maurice] had as much information as we had. He says now he picked up the point in Chargaff's article [the 1:1 base pairing ratio] . . . but he didn't *see* it, and that's all there is to it. Meanwhile Rosalind was wasting time with Patterson superposition methods, and that took her off in the wrong direction entirely. I don't know why she did this. I think Luzzati may have advised it. . . . It was a mistake. But absolutely, she'd have got it out sooner or later."

When asked whether he had confidence in the applicability—however slow in practice—of X-ray diffraction methods when Watson appeared to indicate in *The Double Helix* that he, Watson, had none, Crick said, "Jim doesn't know what he's talking about there. He never understood what [Rosalind] was doing, he simply didn't know enough."

In Crick's opinion, Rosalind should have "done model building too," and when asked why he thought she failed to pursue this method, he said that he did not really know, except that "if she lacked anything, it was intuition." He also thought her "very intelligent, very acute," and remarked that "she did very good analysis. Maurice never saw that, and I think Jim picked up his attitude from Maurice, but she had a good, hard, analytical mind, really first-class." But if she possessed intuition, which he doubted, then "perhaps she mistrusted it." This, Crick felt, was very important in the DNA work, where Rosalind was "too convinced about the evidence" which in some cases proved not very good—"you had to disregard some things. If she had intuition, or had listened to it, she could have seen past what looked like a contradiction. Eventually she would have done this, but it would have taken time. And then she didn't know any biology. That held her up. She didn't have any feeling for biology."

Crick was very little acquainted with Rosalind at the time when the DNA work was going on; subsequently he and his wife formed a friendship with her; and after coming to know Rosalind better, he found it "obvious" that Wilkins's opinion of her was "completely wrong. And what Jim put down in his book is all ideas he had from Maurice. Jim never really knew Rosalind, even afterward. And Maurice had very fixed ideas which Jim accepted. I told him they were wrong."

He added, "Maurice says Rosalind got the B form by accident. But I told him it wasn't an accident he'd managed to have for himself." In Crick's view, this discovery was important, and was far from accidental "except in the way that any successful experiment can be called accidental. You don't predict the outcome, or it's not an experiment, but you don't have a successful result unless you've set it up in a way that allows you to get that result."

He remarked that it was just as reasonable to call the structure that he and Watson did "accidental. . . . Nowadays everyone swears they had powers of prediction, and knew from the outset that the DNA structure would turn out to be significant. But this is arrant nonsense. Nobody knew. . . . This part is luck. There is always a lot of luck in making discoveries. But, of course, you have to be standing in the right place and doing the right thing or you don't get lucky."

When asked for his estimate of the length of time it would have taken Rosalind to arrive at a correct structure for DNA pursuing the methods that she was using, Crick said, "Perhaps three weeks. Three months is likelier. I'd say certainly in three months, but of course that's a guess."

22 Franklin and Gosling, "The structure of sodium thymonucleate fibres. 1. The influence of water content," *Acta Cryst.* 6 (1953): 673.

23 Klug, Aaron, "Rosalind Franklin and the discovery of the structure of DNA," p. 844.

TEN *"What She Touched, She Adorned"*

1 Aaron Klug, on Rosalind's death. The comment is, of course, taken from Samuel Johnson's epitaph on Oliver Goldsmith: "Olivarii Goldsmith, Poetae, Physici, Historici, Qui nullum fere scribendi genus non tetigit—Nullum quod tetigit non ornavit" (June 22, 1776).

2 Rosalind told me this in 1953. She asked rather plaintively, "But how could I stop thinking?"

3 A letter from Maurice Wilkins to Aaron Klug, June 19, 1969.

4 Watson, *The Double Helix*, p. 212.

5 Ibid.

6 Rosalind also told me this in 1953.

[7] There were four Franklin-Gosling publications in addition to the paper which appeared in *Nature*, April 25, 1953. They were:

"The structure of sodium thymonucleate fibres. I. The influence of water content," *Acta Cryst.* 6 (1953): 673.

"The structure of sodium thymonucleate fibres. II. The cylindrically symmetrical Patterson function," *Acta Cryst.* 6 (1953): 678.

"Evidence for a 2-chain helix in crystalline structure of sodium desoyribonucleate," *Nature* 172 (1953): 156.

"The structure of sodium thymonucleate fibres. III: The three-dimensional Patterson function," *Acta Cryst.* 8 (1955): 151.

[8] This was published in English translation in *J. chim. phys.* 50 (1953): C 26.

[9] A letter to the author, December 17, 1953.

[10] A letter to the author, dated Tuesday; the postmark is a smudged date in March 1957.

[11] A letter to the author, dated October 8, 1957.

[12] A letter to the author, December 17, 1953.

[13] A letter from Katarina Kranjc to the author, March 24, 1970.

[14] Recollected from conversation, probably around 1953.

[15] A letter to the author, dated December 20, 1956.

[16] A letter to the author, dated October 8, 1957.

[17] Ibid.

[18] Miss Anita Rimel, for many years secretary to J. D. Bernal, has written about Rosalind,

I found her an extremely, deeply, shy person appearing almost to reject people and so in many cases antagonising them. Without knowing any of the details, I formed the opinion that she had had a bitter struggle at King's College to get proper acknowledgment of her clearly outstanding abilities from the people she worked with. The only person she worked with there with whom she seemed to have friendly relations was Gosling. . . . As a result of this, when she came to us at Birkbeck, she seemed to have armed herself in advance against any repetition of this treatment and tended to be unnecessarily aggressive. I remember teasing her about this on one occasion, and saying that, as a lifelong militant feminist, I well understood the need for fighting for one's rights, but one must not risk alienating oneself entirely in so doing. When I looked at her there were tears in her eyes! I do believe she was badly hurt at King's . . .

I tried very hard to get on friendly, relaxed terms with her and in the end succeeded reasonably well. . . . if only she had known it I was more in awe of her than she could have been shy of me. Her dedication to her research and the clarity of her thinking seemed to me to be on the highest level I knew—that of my Professor's! I secretly thought of her as another Madame Curie, particularly when she persisted, in the face of opposition from the more nervous of her colleagues, in pursuing research into the structure of the polio virus.

This was communicated in a letter to the author, December 1970. It is an interesting view of Rosalind at Birkbeck.

[19] From an interview with Aaron Klug, May 21, 1970.

[20] Ibid.

[21] The chronological list of papers on viruses is:

Franklin, R. E., "Structure of tobacco mosaic virus," *Nature* 175 (1955): 379.

Franklin, R. E., "Structural resemblances between Schramm's repolymerised A-protein and tobacco mosaic virus," *Biochim. et Biophys. Acta* 18 (1955): 313.

Franklin, R. E., and Klug, A., "The splitting of the layer lines in the X-ray fibre diagrams of helical structures: Application to tobacco mosaic virus," *Acta Cryst.* 8 (1955): 777.

Franklin, R. E., and Commoner, B., "X-ray diffraction by an abnormal protein (B8) associated with tobacco mosaic virus," *Nature* 175 (1955): 1076.

Franklin, R. E., "Location of the ribonucleic acid in the tobacco mosaic virus particle," *Nature* 177 (1956): 928.

Franklin, R. E., "X-ray diffraction studies of cucumber virus 4 and three strains of tobacco mosaic virus," *Biochim. et Biophys. Acta* 19 (1956): 403.

Franklin, R. E., and Klug, A., "The nature of the helical groove on the tobacco mosaic virus particle," *Biochim. et Biophys. Acta* 19 (1956): 403.

Franklin, R. E., and Holmes, K. C., "The helical arrangement of the protein sub-units in tobacco mosaic virus," *Biochim. et Biophys. Acta* 21 (1956): 405.

Franklin, R. E.; Klug, A.; and Holmes, K. C. "X-ray diffraction studies of the structure and morphology of tobacco mosaic virus," CIBA Foundation Symposium on the Nature of Viruses, Churchill, London (1956), p. 39.

Franklin, R. E., "X-ray diffraction studies of the structure of the protein in tobacco mosaic virus," *Symposium on Protein Structure*, A. Neuberger, ed. (London: Methuen, 1957), p. 271.

Klug, A.; Finch, J. T.; and Franklin, R. E., "Structure of turnip yellow mosaic virus," *Nature* 170 (1957): 683.

Klug, A.; Finch, J. T.; and Franklin, R. E., "The structure of turnip yellow mosaic virus: X-ray diffraction studies," *Biochim. et Biophys. Acta* 25 (1957): 242.

Klug, A., and Franklin, R. E., "The reaggregation of the A-protein of tobacco mosaic virus," *Biochim. et Biophys. Acta* 23 (1957): 199.

Franklin, R. E. and Holmes, K. C., "Tobacco mosaic virus: An application of the method of isomorphous replacement to the determination of the helical parameters and radial density distribution," *Acta Cryst.* 11 (1958): 213.

* Franklin, R. E.; Klug, A.; and Holmes, K. C., "On the structure of some ribonucleoprotein particles," *Trans. Farad. Soc.* 25 (1958): 197.

* Klug, A., and Franklin, R. E., "Order-disorder transitions in the structures containing helical molecules," *Trans. Farad. Soc.* 25 (1958): 104.

* Klug, A.; Franklin, R. E.; and Humphreys-Owen, S. P. F., "The Crystal structure of Tipula iridescent virus as determined by Bragg reflection of visible light," *Biochim. et Biophys. Acta* 32 (1959): 203.

* These papers appeared after Rosalind's death.

[22] Watson, *The Double Helix*, pp. 225–226.
[23] *Nature* 182 (1958): 154.

ELEVEN *The Last Chapter*

[1] Letter to the author, dated 25 October 1956.
[2] Letter to the author, dated 8 October 1956.
[3] Her mother has written of this occasion,

She invited us . . . to one of the Friday evening lectures at the Royal Institution. The lecturer on this occasion was Professor Robley Williams of the virus laboratory, University of California, Berkeley, U.S.A., and the subject, virus structure. Rosalind herself had lectured at Berkeley and worked with Robley Williams.

In the gallery of the Royal Institution, where the guests were received, photographs and models illustrating virus structure—the work of the American—were exhibited on one side of the room, and on the other stood two large models made by Dr. Franklin and Dr. Klug. In the course of his lecture Professor Williams referred several times to the valuable research and discoveries of Dr. Franklin.

Rosalind was happy that night, gay and vivacious, her eyes sparkling. She wore a red Chinese silk evening blouse that suited her perfectly. It was in the period after her first operations, when her health seemed to have been restored, and she could live a full, busy life with all her old eagerness and zest" (Franklin, Muriel, *Rosalind* [privately published], p. 21).

[4] Letter to the author dated 8 October 1957.

[5] This rather trivial information has been included for a reason. On 19 August 1970 I had an interesting interview with James D. Watson at Cold Spring Harbor, New York. I reproduce here an excerpt from the tape, quoting some of his views about Rosalind:

[Watson] Certainly there was enormous family unpleasantness. [Rosalind's] whole life was tortured. . . . They think differently now. I think it was pretty grim. They didn't want her to go to Cambridge. . . . Her father wanted her to do volunteer social work—she was a very rich Jewish girl. She didn't want any of it. . . . The family seen from the outside is an introspective, self-tortured family. . . . I think it must have been pretty bad. . . . When she had her first operation for cancer she went to recuperate in Francis's house rather than in her family's. . . . The family situation . . . was rather sick . . . worse than most. . . . I suspect the key to most of Rosalind's unhappiness . . . lies with her father.

[After he had seen the manuscript] I had a rather hysterical reaction from Colin Franklin. . . . Then one of

Rosalind's sisters-in-law [Mrs. Colin Franklin] wrote a very nice letter. . . . It led me to put in the epilogue at the end. . . . One should have compassion for Rosalind's environment.

This view of the Franklin family, and their family life, is not only ludicrous, but wholly wrong. Nor is there any belief on the part of those who knew Rosalind well, a number which includes myself but does not include Watson, that Rosalind was a "tortured" person suffering from a deep, chronic "unhappiness." She was, of course, unhappy about things which objectively required that response. Almost anyone would have been miserable working in the atmosphere which surrounded her at King's College. But her personality was fundamentally buoyant, energetic, vital, and optimistic. She got an enormous amount of sheer fun out of life. She was certainly sensitive and rather shy in some circumstances, but her sensitivity was far from abnormal, and her shyness not exaggerated. That her manner with strangers was usually reserved is only a testimonial to a very good English upbringing: in the years when Rosalind was growing up, well-bred English girls were reserved with strangers, and if Rosalind had any slight quality of more-than-usually polite reserve, I think it was only of the kind common to very intelligent people with intellectual preoccupations who do not shift easily into small talk. With anyone she knew well, and liked, she was warm, outgoing, enthusiastic, affectionate, amusing, cheerful, and overwhelmingly thoughtful.

Dr. Mair Livingstone accounts well enough for Rosalind's unwillingness, when she was mortally ill and knew that she was, to return to the parental home. This was not hostility; this was the reluctance of a fatally stricken person fighting with vast courage for a last few months of "normal" living to contemplate the helpless sorrow of those who loved her, and whom she loved.

Watson had no acquaintance with the Franklin family, apart from Rosalind, at the time when he wrote *The Double Helix*. He had not, as I had, dined with Rosalind in her flat in the company of her brothers, her sisters-in-law, her nephews and nieces. Nor had he visited in her parents' home. Rosalind was not like the rest of her family in that she had an extraordinary, highly developed, and specialized set of gifts which they did not share. But a great deal of the vigor, and stubbornness, and enthusiasm, and energy, and dedication, and optimism, and capacity for enjoyment, and thoughtfulness in her character was directly derived, and obviously derived, from her background. So was the aggressiveness of which she was capable, and the argumentativeness. The family as seen from inside by one who knows them is the opposite of "introspective" or "self-tortured"; indeed, they are cultured, lively, witty, involved, outgoing, generous people with many commitments and the native energy to deal with them very competently.

Evidently Watson cherishes a theory about Rosalind which has been very useful in his construction of a consistent, but otherwise quite fictional, character for *The Double Helix*. It is my opinion that the construction of such theories in the absence of acquaintance with, or knowledge of, the facts is a singular impertinence. Scientists, of all people, should know the dangers, and the inadvisability, of leaping to convenient conclusions on the basis of insufficient data.

6 Letter from Dr. Mair Livingstone to the author, 2 March 1973.

7 Ibid.

8 Franklin, Muriel, *Rosalind*, p. 22.

Afterword

1 J. D. Bernal in *The Lodestone* 55, no. 3, Birkbeck College (1965): 41.

2 *Modern Men of Science* (New York: McGraw-Hill, 1966–1968).

3 Dr. Mair Livingstone observed on a visit to the British Museum (Natural History section) that Rosalind's name did not appear in connection with the model of DNA on exhibit there, and wrote to inquire of the Museum's secretary concerning its omission. The reply was to the effect that the Museum made a policy of omitting all acknowledgments, though in fact in this case Watson, Crick, and Wilkins were credited with having done the research on which the model was based. After some further correspondence, Rosalind's name was added in its proper place.

4 Pauling, Linus, "Molecular basis of biological specificity," *Nature* 248 (26 April 1974): 769 ff. The references from which Rosalind is omitted are on page 771.

5 From "Kitty Hawk" from *The Poetry of Robert Frost* edited by Edward Connery Lathem. Copyright © 1956, 1962 by Robert Frost. Copyright © 1969 by Holt, Rinehart and Winston, Inc. Reprinted by permission of Holt, Rinehart and Winston, Publishers.

6 Watson, *The Double Helix*, p. 226.

7 From an interview with André Lwoff, in Paris, 20 October 1970.

[8] Watson admitted this in an interview with me, covered in note 5 to the previous chapter. According to both Aaron Klug and Francis Crick, they each pressed upon him the necessity for adding something to rectify the picture of Rosalind as it stood in the original manuscript. It seems significant to me that such pressure was needed, that in Watson's mind nothing more was required.

[9] The rumor exists that the original title of *The Double Helix* was to be *Honest Jim*. I have no idea whether this is true.

[10] Lwoff, André, *Scientific American* 219, no. 1 (July 1968): 133.

[11] There seems to me some evidence that Watson's attitude toward Rosalind has not materially changed, epilogue or no epilogue. In my interview with him he referred to her several times as "impossible," and "stubborn." He did not make precise the nature of her impossibility," or define the "stubbornness." One gathers he is impenitent.

[12] I conducted a private poll of graduate students at Stony Brook University on various occasions during late 1969 and the spring of 1970. I talked to a small sample, sixteen in all. Nine had read *The Double Helix*. The quotations are accurate; those interviewed were reasonably shy of being named.

[13] Janeway, Elizabeth, *Man's World, Woman's Place: A Study in Social Mythology* (New York: William Morrow, 1971), p. 102.

[14] See *TIME*, 19 April 1971.

[15] Chargaff, Erwin, "Building the tower of babble," *Nature* 248 (1972): 778.